Physicochemical Anthropology
Part II: Comparative Morphology and Behavior

Physicochemical Anthropology

Part II: Comparative Morphology and Behavior

Norman R. Joseph

University of Illinois at the Medical Center, Chicago, Ill.

3 figures and 7 tables, 1979

S. Karger · Basel · München · Paris · London · New York · Sydney

By the same author

Physical Chemistry of Aging. Interdisciplinary Topics in Gerontology, Vol. 8
X + 112 p., 4 fig., 23 tab., 1971. ISBN 3−8055−1084−5 (out of print)

Comparative Physical Biology
X + 234 p., 7 fig., 45 tab., 1973. ISBN 3−8055−1485−9

Physicochemical Anthropology. Part I: Human Behavioral Structure
X + 162 p., 1 fig., 3 tab., 1978. ISBN 3−8055−2793−4

National Library of Medicine Cataloging in Publication
Joseph, Norman R.
Physicochemical anthropology / Norman R. Joseph. − Basel; New York: Karger, 1978
Contents: pt. 1. Human behavioral structure. − pt. 2. Comparative morphology and
behavior.
1. Anthropology   2. Chemistry, Physical   I. Title
GN 24 J83p 1978
ISBN 3−8055−2951−1

© Copyright 1979 by S. Karger AG, 4011 Basel (Switzerland), Arnold-Böcklin-Strasse.25
Printed in Switzerland by Thür AG Offsetdruck, Pratteln
ISBN 3−8055−2951−1

Contents

Contents

Biological Behavior

'All the world's a stage' — W. Shakespeare (As You Like It)
'The Play's the Thing' — W. Shakespeare (Hamlet)

Preface

The behavior of all animals is inseparable from their morphology. This in turn is inseparable from species, age, growth and development. These are the constraining conditions of behavior or ethology in the life of any individual. They also define the limitations of behavior at any period and at any place on the surface of the earth or in its biosphere.

The physicochemical and morphological properties of any animal or species are inseparable or *colligative*. In any non-living aqueous solution, according to thermodynamics and physical chemistry, the colligative properties are those that can be inferred from the properties of water, acting as a solvent for any number of substances dissolved in a system with perfect mixing. In biological systems, water is not miscible within many of the heterogeneous phases. Many of the cells and tissues cannot be characterized as homogeneous solutions with the property of perfect mixing. Nevertheless, such systems may be characterized as showing constant chemical potentials of water and electrolytes at all stages of phylogenetic and ontogenetic development.

In such systems, many properties can be inferred from the chemical potential of water and the electrolyte composition of any type of cellular or extracellular structure, including the homogeneous fluids of the *milieu intérieur*. These colligative properties include muscular tension and work, electrical potentials, and respiratory metabolism. Behavioral processes are colligative in that they involve coordinated changes in all properties that depend on changes of state of the chemical morphology in any part of the neuromuscular system. These behavioral responses are constrained by the normal physicochemical state of chemical morphology as it depends on growth, development and aging of any species at any period of its evolutionary history (*Joseph*, 1971a, b, 1973). These processes can be referred to changes of state of water in the active cells and tissues with resultant changes in behavior and metabolism. These changes should be classed as colligative or 'synechistic'.

The term 'synechism' is derived from the Greek language and means a property of being 'held together'. It can be applied to highly organized and purposeful organisms, as they develop ontologically and phylogenetically over long periods of geological time. The term can also be applied to the continuity of anatomical structure, genes and protoplasm within the life of any individual or species as long as genetic mutations are not free to occur. Insofar as biological behavior is related genetically to species, it is synechistic and continuous with the chemical morphology of any species.

Animal behavior is always dualistic in that it depends on given states of environmental entities and processes. It is always characterized by well-ordered states of reciprocal 'fitness' (*Henderson,* 1908). When sets of biological processes involve coordinated sets of processes in the environment, they may be characterized by the term 'well-ordered or purposeful behavior'. This always implies well-ordered and purposeful changes in protoplasm at cellular and submicroscopic levels. Thus the internal and external dualistic sets of processes are brought into well-ordered and purposeful relationship by a monistic set of behavioral processes. Normal purposeful animal or human behavior is organized and goal-directed by internal and external sets of processes that occur in the living and non-living world by the total process (P) which is essentially monistic rather than dualistic. Coordinated purposeful behavior is integrated with environmental processes through the agency of the sense organs. According to the principles of *emergent evolution* (chapter 1), the behavior and morphology cannot be reduced by Cartesian or Newtonian principles to their mechanistic components in any sense of *identification.* The emergent properties are *well-ordered* and purposeful because of the operation of the laws of thermodynamics and genetics.

In terms of physical chemistry, biological processes can be understood in relation to changes in the dielectric properties of water in the various heterogeneous phases of cells and tissues. These depend on processes of growth, development, aging and ethological behavior during the life of any individual or species. Thus behavior is phylogenetic and ontogenetic. It is constrained by the same thermodynamic, genetic and environmental factors that channel the morphological development of cells and tissues in all individuals and species. Behavior is also subject to the same conditions of 'fitness' that are factors in the adaptation of all cells and tissues to each other, as well as to the given conditions of the external physical, chemical and biological world.

The principle that biological behavior as well as chemical morphology are phylogenetic and adaptive may be expressed as an analytical or deductive syllogism (chapter 2). Thus: all B is M; all M is G; therefore all B is G, where M stands for morphology, G stands for genetics and species, and B stands for behavior. When according to the Henderson's principle of 'fitness', morphology M and E are reciprocally 'fit', then B and E are also mutually fit. Thus both behavior and morphology are mutually adapted to the general conditions of order and disorder prevailing in the environment at any period of genetic and geological history. Adaptive behavior thus depends on the 'fitness' of the sense organs and of neuromuscular responses, which are purposeful, physiological, and genetic.

Introduction

The act of writing is in itself an example of human behavior; it is also a particular case of 'biological behavior'. In former times, let us say in the 18th or early 19th century, the writing would have been performed with pen and ink. In the present century, this is also the usual method of preparing a first draft; the final draft is usually typewritten. Thus the preparation of any manuscript or typescript at any time is carried out by a method that depends on mechanical or technological advances proper to the art of writing. In the case of the typewriter, one expects continuous advances in the machine itself, in the ribbon, and even in the methods of making erasures or corrections.

The early typewriters, developed in the 19th century, were designed to be operated by human beings of either sex. The machine was therefore developed to meet the anatomical and morphological requirements of the adult human body — the size, shape and strength of the ten fingers, the articulation of hands, wrists, elbows and shoulders and the position of the head, forehead and eyes. The alphabetical letters, the punctuation marks, and numbers were also positioned to conform to the normal requirements of the language and to the needs of the average operator. Subsequent developments of the machine were adapted to the varying needs of the market — commercial or personal.

Thus the operation of such an instrument as the typewriter of commerce depends on two fundamental needs. It must be adapted to the requirements of the language as used both by the reader and by the writer. It must also conform to the anatomical and morphological requirements of the average intelligent typist. The technique should preferably be easily learned by any typist of normal intelligence. We may call the first type of requirement — the final use of the product or typescript — the external factor or the factor of final utility. It must be well-ordered, purposeful, and must conform to human value judgements. The other requirement in the actual preparation of a manuscript depends on an internal factor. The instrument must conform to the normal operations of the human neuromuscular system and to the sense organs of the user. Thus biological behavior in general depends on both internal and external conditions of what *Whitehead* (1929) has called *givenness* (chapter 11).

These conditions depend on the states of order or disorder in both the writer and in the potential readers of the manuscript. Thus if the potential readers can understand only one language, French or German, for example, it may be difficult for a writer who understands only English to communicate effectively, using a typewriter of American manufacture, with the American keyboard. Only when the potential readers are proficient in the English language is it possible for the proper states of 'givenness' to be reciprocally well-ordered.

Thus animal behavior, in general, and human behavior in particular depend on reciprocal states of order or 'givenness' in both the organism and in the environment, which includes the typewriter itself as an element or entity. Other elements enter into the performance — for example the illumination of the room. Most people are unable to type in a dark room, although this is not inconceivable. Other conditions of 'givenness' in the environment may affect the performance. Thus adverse conditions of temperature, humidity, or environmental noise, may affect either the physiological efficiency of the typist or the mechanical efficiency of the machine. In general, physiological performance depends on the given conditions in both organism and environment. The human act of walking up a hill, for example is affected not only by the steepness, height and topography of the hill, but also by the physical condition of the walker's legs, ankles, feet and respiration. Every such performance depends on the total conditions of order, disorder and 'givenness' in both organism and environment.

Primary Functions

The primary function of any cell, tissue or organism is to exist. This presupposes a phylogenetic origin of the entire organism, either animal or plant. It also presupposes behavioral relationships with all other organisms in any community of animals and plants within which it shares mutual conditions of life. All animal life implies behavior. Forms of behavior must conform to certain elementary needs of all animals and plants, as conditions for survival of the individual organism and for the species. These needs include those of nutrition, respiration, reproduction, defense, and for symbiotic, predatory and parasitic relationships with other organisms. These needs are shared by all animals at the level of *vie constante:* mammals and birds. At the level of civilized man, the needs must be expanded to include those necessary to meet all the requirements of the economic, social and cultural life of modern nations. The vital activities of modern civilized nations should be reasonable, purposeful, and dependent on a great diversity of skilled behavioral processes that are goal-directed by value judgements common to entire communities and nations. Thus all human behavior can be referred to external factors, as well as to the nature of human morphology and the organization of human protoplasm. We may call morphology and species the primary function of any human being. This function establishes the conditions of constraint of human behavior, both of the individual and of the species. All other functions are therefore secondary and derivative. Properties such as cellular metabolism and respiration are secondary rather than primary (*Joseph,* 1973).

Environmental conditions such as temperature, climate, sunlight and length of the day at any latitude, longitude, or season of the year are constraining conditions of human and animal life. They are among a large number of *external* conditions of 'givenness'. Thus metabolic and behavioral processes at the cellular, microscopic and submicroscopic levels of human or animal morphology cannot be independent of either internal or external states of order, disorder and 'givenness'. A given process in skeletal muscle, for example, depends on states of order or disorder in all the other cells and tissues of the body. It depends on the totality of all conditions in the central and autonomic nervous systems, the cardiovascular and respiratory systems, and on the state of 'givenness' of the entire organism. If all these conditions are exactly fixed, then behavior must be adapted to the state of 'givenness' in the environment. Human neuromuscular behavior obviously is different during a cold day in January (say in Chicago) from that during a hot summer day in New Orleans or in Naples. This is evident in the adaptive behavior of entire populations in these various cities. We can derive few valuable inferences regarding human or animal behavior from the internal physiological, physicochemical or metabolic processes within the intracellular structures, considered in isolation.

As *Bernard* (1878) was the first to point out, internal parenchymal and mesenchymal cells live in a nearly invariant *milieu intérieur.* The constancy of the milieu protects all such living cells from the rigors of direct exposure to the varying conditions of the external physical world *(milieu extérieur).* The living organism does not really exist in the milieu extérieur (the atmosphere if it breathes; salt or fresh water if that is its element) but in the liquid milieu intérieur formed by the circulating organic liquid which surrounds and bathes all the tissue elements: this is the lymph or plasma, the liquid part of the blood which in the higher animals is diffused through the tissues and forms the ensemble of the intercellular liquids and is the basis of all local nutrition and the common factor of all elementary exchanges. A complex organism should be looked upon as an assemblage of simple organisms which are anatomical elements that live in the liquid milieu intérieur.

'The stability of the milieu intérieur is the primary condition for freedom and independence of existence ...'

As a result of the stability and invariance of the milieu intérieur in higher forms of life (vie constante), the intracellular components of the physicochemical system are well isolated from changes of order, disorder or 'givenness' in the milieu extérieur. Stimuli originating in the environment affect internal physiological conditions mainly through the sense organs, those of touch, hearing, vision, taste and smell. Thus decrease of temperature on a cold day in

January is manifested by a sudden change of skin temperature, leading to reflex changes in capillary circulation in the exposed areas. The nature of the weather reaches human consciousness through the howling or whistling of the wind. Auditory reception of the change warns us of the nature of the storm. Human behavior is then adapted in such cases to preserve constancy in the body fluids in intracellular and extracellular phases. Finally, we obtain visual warnings of the external conditions of 'givenness'. Behavior is reasoned and adaptive. It functions to preserve internal states of order and 'givenness'.

It follows therefore that adaptive neuromuscular behavior does not originate in autonomous metabolic reactions within the neurones or myofibrils. These structures are not stimulated by changes within the milieu intérieur, which remains independent of the 'givenness' of external conditions (milieu extérieur). Intracellular components such as adenosine triphosphate (ATP) are not directly responsive to changes of order or disorder in the environment. They respond only to *internal* changes of order within the neurones or myofibrils. These occur through the mediation of the sense organs, which are the only direct contacts of protoplasmic structures with the milieu extérieur. These respond by means of changes of state of the nerve endings in the skin, the ear, the eye, or the other sense organs to induce corresponding changes of state in the various neurones, synapses and centers of the brain and of the central and autonomic nervous systems. Sensory impulses of all kinds are then coordinated to yield motor impulses in the 'final common paths' that ultimately control behavior through the integrative action of the entire body. This reflects an adaptive change of order and disorder in response to changes of 'givenness' in the external world.

In response to protoplasmic changes in the neuromuscular system, there are ensuing changes in respiratory metabolism. These involve chemical reactions involving glucose, glycogen and other nutrient substances. These reactions, involving ATP and other phosphate esters as intermediates, serve to maintain normal *tonus,* irritability, order and homeostasis in the active cells and tissues. Thus the primary processes are kept in control and balance by the secondary processes such as respiratory metabolism, which serves to maintain physiological states of order and 'givenness'. Like respiratory metabolism, physiological and physicochemical processes such as changes of electrical potential, action currents, distributions of muscular tension and length of fibers and ionic transport are secondary rather than primary functions. It is unnecessary for the typist to consider these processes in preparing a manuscript. It is necessary only to consider the given states of the manuscript, the keyboard, and those of his fingers, hands and wrists. It is on these conditions of order that the well-ordered and purposeful skills of the typist depend. All internal physicochemical and metabolic processes at the cellular and submicroscopic levels are secondary functions of the well-ordered morphology of submicroscopic protoplasmic structures (*Frey-Wyssling,* 1953).

Morphology, Behavior and Environment

According to what has been said, all animal and human behavior is an interaction between the chemical morphology or state of order in the organism, and changes in the 'givenness' of the environment. The behavioral response can be initiated either in the organism by voluntary purposeful processes designed to effect changes in the environment, or it can originate in the environment itself. This acts on one or more of the sense organs to produce a new totality of conditions. Symbolically the process may be represented in this way:

$$(M, E) \supset B$$

where (M, E) represents the set of all morphological and environmental changes or processes. Together this set represents the total state of 'givenness' of both organism and environment, which are in continuous states of mutual interrelations. Thus:

$$M \supset E$$

expresses symbolically the principle that internal morphological changes imply correlated changes in the external world. This does not contradict the reciprocal principle that:

$$E \supset M.$$

This expresses the principle that environmental changes may result in changes of chemical morphology of the protoplasm.

In general, the relations are expressed by the diagram shown in figure 1. Here the circle M represents the set of all systems of chemical morphology in the organism. The circle E represents the set of all environmental objects, entities and processes. The shaded area B represents the region in which the sense organs receive stimuli or continuous sense impressions through nerve endings of the skin, eyes, ears, and the organs of taste and olfaction. The area B also represents the region of all processes in which the organism acts on the environment, principally through the neuromuscular system.

The region B would vary in depth and area according to the nature of the species; it depends on whether this belongs to *vie constante* or to *vie oscillante*. In mammals, B would depend on the size of the animal. In small mammals such as guinea pigs, rats or mice, the total surface area is large in relation to the total mass. This is explained by the 'principle of similitude', according to which basal metabolism is proportional to the ratio of S/V, where S is surface area and V is volume (chapter 9). The ratio is high for small animals, and low for species such as men, horses and elephants. The ratio increases by a factor of 10 for each 1,000-fold decrease of mass or volume.

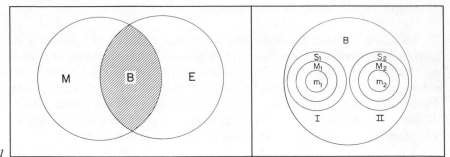

Fig. 1. Relation between morphology (M), behavior (B) and environment (E). M is set of all morphological entities and processes. E is set of all environmental states and processes ('givenness' or 'order'). All behavior is in both M and E. It includes all processes relating to change of givenness or order in (M, E). E affects M through nerve endings in the sense organs. M affects E through neuromotor responses.

Fig. 2. Behavior of two participants in a game. Outer circle B represents class of all behavioral processes in game played by two individuals, 1 and 2. S_1 set of all processes in sense organs of 1; S_2 is the corresponding set in the sense organs of 2. These sets are represented by the outer circles of the concentric circles representing neuromuscular behavior of each individual. Neuromuscular are in S: all S is in B. Therefore M_1 and M_2 are in B. These two sets are not independent but depend on the progress of the game, B. They depend on the processes in the sense organs, which also depend on B. Finally, metabolic or respiratory processes at the cellular or submicroscopic levels are in M and S. Thus they depend on neuromuscular behavior, and are not the sources of energy of muscular contraction, as postulated in the theory of active transport.

Behavior is not in the intracellular metabolic processes, m_1 and m_2. Morphological and metabolic processes are in the behavioral processes as represented by B, the progress of the game. Relation of behavior to metabolic processes is symbolized as:

$$B \supset (S, M, m).$$

This states that the set of all behavioral processes implies the set of all concurrent sets of processes in the sense organs, S, in neuromuscular morphology M, and in intracellular metabolic processes. All S, M and m are in B, as is shown in the figure.

A man weighing 70 kg may have a basal metabolism of about 2,000 kcal/ day. According to the principle of similitude, a small rodent, weighing 70 g, would have a basal metabolism of about 200 kcal/day, or about 3 kcal/g. This value is much greater than the value for man – about 0.3 kcal/g. Thus the nutritional requirements of a mouse are much greater than those of a dog, horse or a man on a weight basis. This compels small animals to consume a much larger quantity of food than that required by large mammals on a weight basis. Accordingly, rodents must be continuously in pursuit of food. This leads to very active muscular movements at all periods, to very rapid motion in relation to size, and to many other effects on behavior. Thus the area B representing behavior must be enlarged and expanded into both M and E, which represent the nutritional needs of the animal with respect to nutritional supplies. In large mammals, such as man, a much smaller fraction of metabolic energy is required to meet the nutritional needs. Thus total behavior of any mammal depends on morphological factors such as species and size. In general, behavior is as much a property of phylogeny and ontogeny as is the total chemical morphology and basal metabolism. These various aspects of mammalian life cannot be isolated or studied as independent aspects of the total life of the organism.

In figure 2, we may represent the behavioral relation between two individuals engaged in mutual or reciprocal activity such as a game. A tennis match, a foot race or a game of cards could serve as examples. The large circle B (behavior) represents the set of all biunivocal processes of the game. The two circles, S_1 and S_2, represent sets of processes in the sense organs of the two individuals; these are the mutual means of communication. The neuromuscular responses are represented by the two circles, M_1 and M_2, which represent the changes of chemical morphology and state of the two players. These circles are in S_1 and S_2, representing sensory communication. The progress of the game, as represented by B, corresponds to the simultaneous sets of events, M_1 and M_2 in the two neuromuscular systems. Finally, the intracellular metabolic processes in S_1, S_2, M_1 and M_2 correspond to the sets of metabolic processes, m_1 and m_2, in the two individuals. These also correspond synchronously with the progress of the game. Games involving more than two players such as basketball or soccer could be represented by extending the number of circles to represent all the competing players as well as the referee, scorekeeper and other officials. The behavior of a large crowd could be thought of as the resultant changes in S, M and m for each individual.

Transfer of Order

In the normal human act of walking, each leg is alternately contracted and extended; this implies an alternating transfer of order and disorder from one leg to the other. In the extended state of any group of muscles or muscle fibers, the intracellular water is in a well-ordered state of low dielectric constant and high dielectric energy (*Joseph*, 1973). As the muscle is contracted, the dielectric constant increases, and the state of water becomes relatively disordered. Glycolysis is stimulated, and dielectric energy (configurational free energy) is converted to work. Each act of contraction implies a conversion of configurational free energy to work. This is given by:

$$\Delta G_c = -T \Delta S_c = -W_{max}$$

where ΔG_c is the free energy change, ΔS_c is the change of configurational entropy, and W is the maximal external work. Repolarization of the fibers requires a reversal of the process that depends on a supply of free energy or negative entropy from nutrient sources of glucose or glycogen. This implies a reordering of the muscle fibers to the well-ordered state. If this is denoted as O, and if the contracted state of disorder is denoted as D, then the process of bipedal walking may be indicated as:

Left leg: O   D   O   D   O   D   O   D   O   D   etc.
Right leg: D   O   D   O   D   O   D   O   D   O

Thus order is transferred from one leg to the other with each stride. Each leg is alternately in the well-ordered state or in the disordered state. If the transition from one state to the other is not taken into account, then the probability of either leg being in the state O is one-half. In a more complex process of muscular activity that involves many muscles, each of which can exist in either of two states, a case may be considered in which the probability of any muscle being in state O is exactly one-half at any time. If the configurational free energy or configurational entropy of a set of n such muscles remains constant, the system does no external work and may be said to be 'conservative'. This would be the case in a human subject if the fingers, hands and arms were free to transfer order and disorder alternately from the right side to the left with no performance of external work. If this were assumed to involve a total of n muscles, each of which is permitted to exist in two different states, then the total number of possible configurations would be of the order of 2n, where n denotes only *possible* changes of state. Since the probability of the ordered state is one-half for each muscle, the number of *possible* configurations for the entire set is:

$$N = 2^n$$

Thus the number of possible configurations of the body in a state of constant free energy tends to increase exponentially with n, the set of all muscles that can change configuration or position. This implies a very large number of possible positions in states of constant free energy, entropy and constant metabolism.

If the system is *nonconservative,* external work is performed at the expense of nutrient metabolic energy. The exchange of energy is governed by the condition:

$$\Delta G = \Delta H - T\Delta S$$

where G denotes free energy, H denotes enthalpy, and S stands for entropy. The conversion of glucose to lactic acid is a process in which $T\Delta S$ is 12.21 kcal/mole of glucose. Correspondingly, ΔG is -29.88 kcal, and ΔH is -17.67 kcal. Thus in a *nonconservative* system the loss of free energy ($-\Delta G$) exceeds the loss of heat ($-\Delta H$) by 12.21 kcal/mole of glucose. This is the amount of free energy ($T\Delta S$) required to maintain constant muscular tone, irritability and 'givenness' in each of the alternate returns of the fibers to the extended states. Thus in walking, each extension of the leg requires an expenditure of metabolic energy. This is in excess of the normal supply in the conservative or resting state. Thus any behavioral activity requires rather complex exchanges or transfers of free energy, entropy and dielectric energy from one set of muscles to another with concomitant changes of respiratory metabolism. This is stimulated in the disordered state of high dielectric constant, and inhibited in the well-ordered state. Normal neuromuscular behavior then implies labile and reversible changes of state of intracellular water, related to processes of electrical polarization and depolarization and to changes of respiratory metabolism. These intracellular processes are secondary. The primary process is the well-ordered purposeful behavior in which the changes of chemical morphology in all the anatomical structures at every level are coordinated. Thus the act of using a typewriter is well-ordered, purposeful and primary. The concurrent intracellular processes in the neuromuscular system are secondary to the primary process, which is the systematic well-ordered change of chemical morphology at the cellular level. The primary and secondary processes are coordinated by the sense organs and brain, which operate to bring the well-ordered internal processes into relation with the varying states of order and disorder in the external world. This involves all the purposeful and well-ordered activities of the human body, including metabolic control of homeostasis and 'givenness'.

The foregoing considerations show that all behavioral processes are dualistic in that both internal and external changes are coordinated by a kind of 'mirror image' correspondence. If we follow *Whitehead*'s (1929) conceptions, this can be expressed by the relation:

$$P \supset (M, E)$$

where P (process) implies the set of all morphological and environmental changes. These represent respectively the internal and external aspects of behavior as coordinated by the sense organs. These serve to unify the internal and external worlds and to yield purposeful relationships. Thus (M, E) is a dualistic set: P is the unifying principle, which is 'monistic'. When in the common sense view of the world, people behave more or less instinctively, they are monists. They are instinctively applying *Occam*'s razor by reducing a dualistic set of entities to one set, which may be termed 'process'. Excessive cerebration tends

to defeat *Occam* by yielding to Cartesian reductionism and mechanistic determinism. These are essentially nominalistic and particularistic in nature. Instinctive behavior may operate to favor the liberation of the general, the
universal, the poetic and the generally pleasurable aspects of life. 'Process' is a
concept that is opposed to 'mechanism' and to 'reductionism'. If informed by
natural, instinctive forms of rational behavior that promotes the art of living, it
may in the long run serve a liberating function.

Human and animal behavior may involve two or more individuals of one or
more species. The behavior of groups of organisms is integrated through sets of
processes of the form:

$$(N_1, M_1, m_1 \ldots) \rightarrow (N_1', M_1', m_1' \ldots)$$
$$(N_2, M_2, m_2 \ldots) \rightarrow (N_2', M_2', m_2' \ldots)$$

where N_1 refers to sense organs and to nerve, M_1 refers to muscle and m_1
denotes metabolism. $(N_1', M_1', m_1' \ldots)$ refers to the changes of state of the set
S_1, denoting the set of all subsets.

Then

$$S_1 \rightarrow S_1'$$

and

$$S_2 \rightarrow S_2'$$

These are coupled or conjugated in the following way:

$$(S_1, S_2,) \rightarrow (S_1', S_2')$$

where the changes of S_1 and S_2 are denoted as S_1' and S_2'. These are reciprocally
related in all mutual behavior — for example in games, contests, work, verbal
discourse or social relations. The process P is then of the nature:

$$P \supset (S_1, S_2)$$

where S_1 and S_2 denote two sets of behavioral responses. The main components
of each set include the sense organs and neuromuscular systems. These cannot be
reduced mechanistically to the individual components considered in isolation.

Chapter 1

Reason, Skill and Value

Like *le bon sens* of *Descartes,* reason is the most widely distributed of human qualities. No human being seems to be disturbed by his deficiencies in common sense, however conscious he may be of his other failings. But people differ greatly in their evaluations of human qualitites. They have different aims and goals, and what is important to one man may be meaningless to succeeding generations. The search for permanent and universal values has been mainly confined to philosophers and metaphysicians. There is a unity of philosophical experience based on the quest for Reason and Value. The fundamental questions raised by *Plato* and *Aristotle* have persisted through the ages down to the present period of *Whitehead, Russell* and *Alexander.*

Reason has been described by *Whitehead* (1929) as follows: 'The function of Reason is to promote the art of life.' This would seem to be a more universal description than to relate it to the *science* of life. Many people who may be completely indifferent to the values of science or to a science of life, are vitally interested in reason, or the art of life. These cannot be described as unreasonable people. In fact, their attitude was the only conceivable one before the dawn of modern science in 17th century Europe. At all other places and at all other times, with the possible exception of certain periods in the life of Ancient Greece, this was the only possible attitude.

Pascal in his *Pensées* has described the two extremes of the life of reason, in contrasting 'l'esprit de géométrie' with 'l'esprit de finesse'. The geometrical spirit seeks to reduce life and nature to exact logic and mathematical laws. The spirit of finesse sees life as perceived by the human senses in direct communication with nature and with man. It is the approach of the poets, the artists and in science the approach of the naturalists. At its highest level, the life of art and the life of science were shared by universal geniuses such as *Leonardo* and *Michelangelo.* Their greatness ranged from painting, architecture, poetry, engineering and sculpture *(Michelangelo)* to painting, technical inventions, physiology, anatomy and animal locomotion *(Leonardo).*

Attainments at this level demand not only a quest for ultimate values, but also the development of the highest skills. Good intentions are not sufficient.

Disciplined training of the mind, the hand, the eye, and of all the senses are demanded. Value judgements are required at every step in the life of the mind. Human values, for their attainment, then require creative abilities that employ all the skills available from the special sciences added to those of common sense and qualities of judgement.

Physiology of Skill

'By no means every physicochemical process represents a life process, though every event in a living organism is a physicochemical phenomenon' (*Lorenz,* 1970). Thus the behavior of human beings and other higher organisms may be regarded in every case as a well-ordered pattern of physicochemical processes that involve the neuromuscular system, the sense organs, the brain and the systems of intracellular respiratory metabolism. Any physicochemical change of state, or biological process, may thus involve many sets of morphological and metabolic phenomena that depend on changes of state of aggregation of intracellular and extracellular sets of colloids, electrolytes and miscible or immiscible phases of water (*Joseph,* 1971, a, b; 1973). In the following chapters, it will be shown that these biological behavior patterns involve the organization of the central nervous system, the various sense organs, and central or spinal reflex arcs. When, as in creative human behavior, reason, skill and value judgements are involved, the highest forms of behavioral processes depend ultimately on the physico-chemical organization of all cells and tissues.

According to *Frey-Wyssling* (1953), '... orderly biological processes are unthinkable without presupposing structure, and it is therefore out of the question that any living constituent of protoplasm could consist of structureless fluid, or of independently displaceable particles'. It would follow from this opinion that orderly processes of biological activity, including all creativity that involves reason, skill, and value judgements, imply well-ordered submicroscopic structures in living protoplasm; they cannot be explained merely by Brownian movement or changes of chemical bond energy in homogeneous solutions that lack *purposeful* organization.

It would also follow from the views of *Frey-Wyssling* and other morphologists that, if the chemical morphology of cells and tissues has developed through adaptive ontogenetic and phylogenetic processes, the evolution of biological behavior in the world of living organisms would follow laws of population genetics. The evolution of purposeful behavior would then depend on the phylogenetic development of any given species. Ultimately, the evolution of behavior is limited by the same conditions of contraint that channel the physicochemical development of chemical morphology of any species. These constraints are of two kinds — genetic and thermodynamic or physicochemical (*Joseph,* 1973).

Highly developed, well-ordered and purposeful sets of human activities may be denoted as skills. As stated above, they are based on *reason* and *value* judgements that groups of human beings consider to promote in any way the art of life.

Ultimately these skills have a physiological and physicochemical basis in the organization of protoplasm in the human brain, neuromuscular system and in the various sense organs and receptors. Any skill depends on highly organized, purposeful behavior patterns that cannot be limited to one specific form of activity. Any skill worthy of the name engages all the activities of the mind and body. To attempt to reduce it geometrically to its biological components is to disintegrate its character of 'agapism', and to devalue it to a mere cybernetic mechanism.

The term 'agapism' is here used in the sense applied by *Peirce* (1923) to that quality of life based on love, the affections or emotions that differentiate the living animal from a physicochemical mechanism or automaton. According to what has been said, this must be based on the evolutionary development of reason as the peculiarly human form of the art of living and cannot be independent of the orderly development of submicroscopic structures of protoplasm.

As an early attempt to place the art of living on a physiological or scientific basis, one should take account of the famous 18th century treatise on gastronomy, *The Physiology of Taste* by *Brillat-Savarin.* This work eminently satisfies *Whitehead*'s criteria as to the functions of reason: to live, to live well, and to live better. As everyone knows, the first function of absorbing food in the diet is to live. Any person who has survived to a mature age in any state of body or mind has met the first condition. Only those who have met the third criterion — to live better — would have satisfied *Brillat-Savarin*'s conditions for gastronomy as an art of life. These conditions call for the arts of cuisine and the use of fermented or distilled liquors to meet strict criteria, not only for the satisfaction of the senses, but also for intellectual pleasures and for sound mental and physical health. Gastronomy, so defined, satisfies the human needs for reason, skill and value judgements.

As a man of the 18th century enlightenment, *Brillat-Savarin* went far beyond the standards of later authorities on nutrition, who have tended to adopt a more utilitarian view. Especially in the more pragmatic minds of 20th century nutritionists, a satisfactory human diet must consider only the caloric values and the necessary accessory factors, such as the vitamins. In accordance with the principles of a Cartesian mechanistic science, 'agapism' or the pleasure principle is often regarded as of secondary importance. Thus only the first of *Whitehead*'s conditions for the art of living is satisfied — the need to survive.

Agapism

Three modes of evolution in the universe have been described by *Peirce* in an essay on *Evolutionary Love* (1893, 1955). These refer to evolution in general, and are not limited to organic or biological evolution. The modes are distinguished as evolution by fortuitous variation, by mechanical necessity and by creative love. '... The mere proposition that absolute chance, mechanical necessity and the law of love are severally operative in the cosmos may receive the names of *tychism, agapism* and *synechism.*' Elsewhere *Peirce* uses the term synechism as applied to mechanical necessity, laws of causality or the principle of continuity. The concepts of tychism and synechism will later be defined more explicitly in chapter 2, which deals with logical concepts. All three principles apply to reason, the art of living and the use of value judgements and skill.

The English language, especially in recent and contemporary usage, is rather impoverished in providing adequate synonyms for terms such as *agapism* or evolutionary love. To find adequate verbal expressions, *Peirce* and other modern writers have found it necessary to use terms derived from ancient mythology, such as Greek, Latin, Hebrew or Egyptian sources. Agapism may thus be used in the sense of Eros, the god of love, or the goddess Venus, who has always symbolized the power of creative love or evolutionary love. In the period of the High Renaissance in Italy, the power of creative love is shown in famous paintings of the Madonna by *Leonardo, Michelangelo,* and *Raphael.* In his chapter on *The Dynamo and the Virgin, Adams* (1900) described the power of creative love, as represented by the Madonna as the driving force in the creation of the great cathedrals and religious art of Medieval Europe. In the 20th century, he saw that force displaced by the dynamo, which represented the creative power of physical energy. Thus, *agapism,* in human civilization, was by that time largely replaced by *anancism,* the power of evolution by mechanical force.

In referring to views of evolutionary love, *Adams* called attention to the great poem of *Lucretius On the Nature of Things.* In his dedication, *Lucretius* addressed the goddess in this way: 'Mother of Aeneas and his race, delight of men and gods, life-giving Venus, it is your doing that under the wheeling constellations of the sky all nature teems with life, both the sea that buoys up our ships and the earth that yields our food. Through you all living creatures are conceived and come forth to look upon the sunlight.' *Lucretius* continues in this vein for many verses. It is obvious that this view of the creative force was far from 'l'esprit de géométrie', which since the period of *Descartes* and *Newton* has come to lead to a reduction of science to the principles of synechism and tychism, or to chance and logic. This has been the main result of the repression of physiological sensation and perception in the interests of abstract geometrical expression and mathematical or logical formulation. A set of agapastic values has given way to historically modified sets more in harmony with mechanistic and utilitarian values. The process was clearly understood at the time at which *Adams* analyzed the historical forces that developed the power of the dynamo.

Emergent Evolution

The history of the principles of *emergent evolution* has been briefly outlined in a short book on the subject by the entomologist and naturalist *Wheeler* (1928). The principle is of general application and finds illustrative examples in the world of inorganic substances as well as in biology. *Wheeler*, as a student of animal societies, has found the principle useful both in homogeneous societies, consisting of only one species, as well as in heterogeneous communities consisting of many species living in various kinds of associations. These include the *biocoenoses*, or 'communities of plants and animals that live in particular environments, such as swamps, deserts, rain forests, etc. – veritable welters of organisms of many species, all interacting with one another in complex predatory, parasitic and symbiotic relationship, but forming wholes in which the experienced field naturalist can readily distinguish general adaptive patterns ... We may truthfully say that there is not on the planet a single animal or plant that does not live as a member of some biocoenose.'

Among the 20th century American and British philosophers and biologists committed to the principle of emergent evolution, *Wheeler* included the following: *Holt, Spaulding, Sellars, Alexander, C.L. Morgan, Gordon, C.K. Ogden, G.H. Parker* and *Jennings*. The general set of beliefs of this group can be summarized in the words of *Spaulding:* 'Certain specific relations organize parts into wholes ... certain states of affairs are identical with new properties and are different and distinct from the individual parts and their properties. Therefore the *reduction* of these new properties to the parts in the *sense of identification* and the finding of a *causal determination* also in the same sense is *impossible.'* This statement may be compared with one by *Willstatter* (1928).

'Until now it was taken for granted in chemistry that the properties of components in chemical compounds disappear but that in mixtures of substances they are retained. This is an antiquated viewpoint ... Mixtures may actually possess the nature of new chemical compounds.' Although he may have been unaware of the views of *Spaulding, Willstatter* would undoubtedly have agreed that the reduction of the properties of protoplasm to those of its constituent components is *impossible*. According to *Oparin* (1938) '... this compels us, in considering the evolution of organic substance to rely not upon those alterations in which this or another isolated compound may be subjected, but to bear in mind alterations which take place in complex mixtures of various organic substances.'

These views of an eminent biochemist and of a student of the phenomena of life emphasize what may be called an organicist point of view, which is wholly compatible with the views of *Wheeler, Spaulding* and the other biologists that are listed above. The properties of protoplasm are those of an emergent entity, which has evolved from primitive antecedents. Organicist views were also held by *Henderson* (1928) in his treatment of blood as a physicochemical system. This treatment could be considered to depend on the emergent properties of a system containing water, the serum proteins, the erythrocytes and hemoglobin in dynamic equilibrium with physiological ions, with the respiratory gases and with other components. The emergent properties are those of a well-ordered, purposeful physicochemical system which combines the properties of an invariant milieu intérieur (*Bernard,* 1878) with those of a heterogeneous system containing several classes of components and phases in physiological and physicochemical relationship with body cells and tissues and with the lungs. It is impossible to reduce the properties of the physicochemical system to those of its constituent parts in isolation.

Organicist views were also held by *Bernard* (1865, 1927). 'In chemistry, synthesis produces, weight for weight, the same body made up of identical elements combined in the same proportions, but in the case of analyzing and synthesizing the properties of bodies, i.e. synthesizing phenomena, it is much harder. Indeed, the properties of bodies result not merely from the nature and properties of matter, but also from the arrangement of matter. Moreover, as we know, it happens that properties which appear and disappear in synthesis and analysis, cannot be considered as simple addition or pure subtraction of properties of the constituent bodies. Thus, for example, the properties of oxygen and hydrogen do not account for the properties of water, which result, nevertheless, from recombining them.'

'... I shall here only repeat that phenomena merely express the relations of bodies, whence it follows that, by dissociating the parts of a whole, we must make phenomena cease if only because we destroy the relations.'

Finally, *Bernard* concludes these reflections with the observation: 'I am persuaded that the obstacles surrounding the experimental study of physiological phenomena are largely due to difficulties of this kind; for despite their marvelous character and the delicacy of their manifestations, I find it impossible not to include cerebral phenomena, like all other phenomena of living bodies, in the laws of scientific determinism.'

Sets and Aggregates

The fact that the properties of protoplasm are irreducible to those of its inorganic or macromolecular components in the state of homogeneous mixtures is explained by the fact that the properties are those of definite sets of components in definite states of organization. This implies that we are dealing with states of order or disorder; in physical chemistry this signifies that thermodynamic functions such as free energy and entropy are involved.

The properties of aggregates of several kinds of macromolecular components coexisting with water and inorganic electrolytes require for their description the conditions of chemical morphology and physicochemical state (*Joseph*, 1971a, b; 1973). These are the primary properties of any system of cells and tissues, and of the organism as a whole. All other properties are secondary — including all the physicochemical and metabolic properties which depend on chemical composition and physicochemical state of the cells and tissues. Since physicochemical state is an emergent property of chemical composition, it cannot be inferred from analytical data without reference to the physicochemical state of aggregation in the normal or standard physiological state. This primary property is a result of a long process of organic evolution. In each type of cell or tissue it depends on the species (phylogeny) and on the age and state of development (ontogeny) of the entire organism. These are emergent properties that develop according to the principles of tychism, synechism and agapism. Since animal behavior depends also on principles of emergent evolution, it is also a secondary property of chemical morphology and state. Behavior also depends on the place in nature of any given species.

In the state of nature, according to *Wheeler*, every animal flourishes as a member of some kind of biocoenose. It enters into mutual behavioral relationships with all other species within the biocoenose. This includes mutual social relations with other members of its own species. The behavior of the entire community is an emergent of the morphology and behavior of all its members. To understand these aspects of evolutionary behavior, it is necessary to consider the logic of sets, classes and aggregates.

Cantor has defined the concept of aggregates as follows: 'By an aggregate *(Menge)* we are to understand any collection into a whole *(Zusammenfassung zu einem Ganzen)* M of definite and separate objects of our intuition or our thought. These objects are called the "elements" of M.

In signs we express this thus:

$M = (m).$'

In the logic of sets and aggregates, the properties of the aggregate (M) cannot be inferred from the properties of the individual entities (m). Nor is it possible to infer all the properties of the constituent objects (m) from those of the aggregate. Thus the principles of emergent evolution are in agreement with the logic of aggregates as formulated by *Cantor*. They are also in agreement with the classification of types of physicochemical systems as expressed in the thermodynamics of homogeneous solutions or mixtures, or of heterogeneous aggregates of immiscible substances (*Gibbs*, 1875, 1928).

For example, a homogeneous mixture or aqueous solution of several substances can be characterized in at least two ways. Several of its properties may
be treated, as a first approximation, in a general way applicable to any such
aqueous mixture. These are the 'colligative' or nonspecific properties of the
solution, treated as an aggregate. However, as a second approximation, it is
necessary to consider the exact chemical nature of each component in its mutual
relationships with all the other members of the set. Thus the general properties
of physicochemical sets cannot be inferred from the properties of each of the
components. It is necessary to consider the exact chemical nature of each
component in mutual relationships with respect to solubilities, electrical valence
type and chemical interactions. The property of each such homogeneous or
heterogeneous physicochemical system is an emergent of the properties of each
component and phase. Thus the biological principle of emergence in morphology
and behavior is well founded, not only on logical and mathematical grounds, but
also on physicochemical and thermodynamic principles as applied to all homogeneous and heterogeneous systems.

Indeterminacy

The development of quantum mechanics in 20th century physics has had a
very profound effect on concepts of law and causality in many realms of
thought. Together with developments connected with relativity theory, revisions
of physicochemical thought and methods have been everywhere required. This is
also true of questions as to the ultimate meanings of many fundamental physical
observations. The nature of the problem may be illustrated by *Heisenberg*'s
'principle of indeterminacy'. At the atomic or subatomic level, there is a
fundamental uncertainty in determining by any means the exact coordinates q
and the exact momenta p of any particle such as an electron in any state of
quantization. This is expressed by a relation of the form:

$$\Delta p \, \Delta q \geqslant \frac{h}{4\,\pi}$$

The relation is used to determine the sharpness of a spectral line, the
frequency of which is proportional to the difference between two energy levels:
this is given by $h\nu$, where h is *Planck*'s constant, and ν is the frequency of the
emitted radiation. The term Δp is the uncertainty of the corresponding coordinate. Thus as either parameter approaches zero, the other becomes correspondingly large.

Heisenberg's principle and its statistical interpretation have led to great
successes in theoretical physics. But, by its statistical and indeterministic nature,
it has brought about fundamental revisions in the instinctive human desire for

strict causality and law in the ordering of physical and chemical phenomena. Thus fundamental revisions in the approach to scientific causality have been required, and these have not always accorded with traditional ideas of cause and effect. The dilemma has given rise to philosophical problems, the resolution of which compels basic revisions of traditional ideas of organism and mechanism. This has led *Whitehead* (1925, 1929) to advocate a philosophy of organism not only in the biological sciences, but also in the physics of fundamental atomic and subatomic processes. Thus the foundations of mechanism have been challenged in the citadel where it has prevailed for nearly three centuries.

This forces us to give up the fundamental Newtonian principles of particle mechanics, which regard the position of the particles in absolute space and time as the fundamental reality. If the motion of any particle is not absolutely determinate, then the fundamental reality cannot be described by absolute coordinates. The Newtonian conception of motion as a function of force, acceleration and time, must be replaced by the conception of the coordinates and momenta as subordinate to the *event,* which is the fundamental reality. This leads *Whitehead* to the concept of atomic organicism, which could be regarded as an emergent of all the sets of electronic and subatomic particulate energies and motions.

If this view is accepted, then even in the field of atomic physics, it is necessary to give up traditional Newtonian, Cartesian and Laplacian views of mechanistic determinism in favor of 20th century organicism, which gives priority to the emergence of well-ordered aggregates from relatively independent disorganized elementary units. In the biological realm, this emphasizes the fundamental importance of submicroscopic *structure* of protoplasm as the basis of all ordered biological processes (*Frey-Wyssling,* 1953). This points to the importance of the organismic properties of protoplasm as a well-ordered heterogeneous system – an *emergent* of the component organic and inorganic substances that contain water as an integral part of the structure.

Like physical causality, biological causality is then fundamentally statistical in nature. Well-ordered causality tends to be diminished or to disappear under conditions where the fundamental structure is disintegrated toward a state in which the components become disaggregated, disordered and independent.

Biological Processes

In general, two main types of biological processes may be distinguished: very rapid, and very slow or gradual. Each type implies characteristic changes in the chemical morphology of cells and tissues. Rapid processes in man include those that are carried out in the course of the daily activity. These would include walking, running, or other types of muscular activity from which a man recovers to a normal standard state within, at most, a 24-hour period of rest. These processes include all the activities carried out during the day's work, which involve no permanent irreversible changes in the cells and tissues of the body.

The other type of biological process includes the normal processes of growth development and aging. In the growth period of mammals, the cells and tissues change irreversibly, but in large mammals such as man, the changes are relatively slow, especially after the attainment of adult proportions. In early life, especially during the period of rapid formation of bone, many cells and tissues undergo marked changes of chemical morphology, including the normal distributions of colloids, water and electrolytes. Liquid phases, such as blood plasma and other fluids of the milieu intérieur remain almost invariant. Important changes are mainly confined to the solid or semi-solid phases of intracellular structures, in which water occurs in an organized, immiscible state. In most cellular and intracellular phases, the distribution of water and electrolytes does not conform to the theoretical formulations of the classical theory of 'membrane equilibrium', or to the *Nernst* formulation for 'membrane potentials'.

Distributions and potentials are determined by the general conditions for structured immiscible phases (*Gibbs,* 1875, 1928; *Joseph,* 1971a, b; 1973). Distributions of water and electrolytes are primary properties that are determined by the standard chemical morphology of all cells and tissues. They are emergent properties that depend phylogenetically and ontogenetically on the development of each species. In mammals (vie constante) especially, they depend on thermodynamic conditions of invariance and constraint that maintain a constant milieu intérieur. This depends on normal water and electrolyte balance. In the growth period, therefore, mammals develop with one degree of freedom. There is only one normal mode of growth and development, and only one thermodynamic degree of freedom. Growth and development of every kind of cell and tissue therefore depend on two kinds of constraints: thermodynamic and genetic.

Given these normal standard conditions for primary biological processes of growth, development and changes of chemical morphology, there are also changes of the secondary properties that depend on morphology as the primary emergent function. These secondary properties include all physiological and physicochemical functions that depend primarily on morphology. The secondary properties are likewise emergents that depend organismically rather than mechanistically on the development of the whole animal. They include rates of respiratory metabolism, growth rates, reversible and irreversible electrical potentials, muscular tension and work, transport of oxygen, carbon dioxide and heat, as well as nerve conduction and organization. In their character as secondary properties of chemical morphology, these attributes of the organism must be considered to be emergent rather than independent entities or units. Physiological behavior at any age or state of development is also an emergent which depends on these primary and secondary properties of chemical morphology. All these properties are functions of time. From the broadest point of view, they must be considered in relation to the general tychistic, agapastic and synechistic modes of organic evolution.

From the point of view of symbolic formulation of the above relationships, the following notation may be employed. Let S_1 represent the set of all properties that determine the chemical morphology of any set of cellular phases C_1: Let S represent the emergent or organismic set of properties of all subsets (S_1, S_2, ... S_n). Then:

$$S = (S_1, S_2, ... S_n).$$

This set is regarded as primary. Then if S_1' is the set of all secondary properties that correspond to S_1, the chemical morphology of the first kind of cellular or intracellular structure,

$$S_1' = (S_1)$$

and

$$(S_1', S_2' ... S_n') = (S_1, S_2 ... S_n).$$

Accordingly, all the secondary properties, considered as a set, depend on the primary set that describes chemical morphology and physicochemical state. This is true of all the primary and secondary properties with the exception of those that develop in time as mental processes, skills, reason, or values. For a given individual, these are *limited* or *constrained* by the conditions of ontogenetic development, but they are not absolutely *determined* by those conditions.

Thus at any age or state of development, any human being is free to change his mental attitudes, thought processes or abilities. This freedom is certainly limited by previous development, but not in an absolute way. The entire process of education or training depends on this freedom. At any age, the human being can develop new skills or improve his old ones. This development depends on the opportunities afforded by society. The opportunities at any time are not absolutely determined, but depend to a certain extent on chance. The evolution of human skills and the development of reason and value judgements thus depend in part on tychistic factors of human evolution.

Accessible States

The standard state of homeostasis in man has been defined as a state of maximal stability or as the most accessible state (*Joseph,* 1971a, b; 1973). This is a state of basal metabolism in which the rate of energy conversion is minimal and constant. It is also a state of minimal configurational entropy or maximal dielectric energy. In this state, the dielectric energy remains constant in the organism as a whole, but is not necessarily constant in a given tissue or set of tissues. In the resting state, energy and rates of respiratory metabolism are continually redistributed under standard conditions of constraint for the organism as a whole. This is a result of integrative behavior. As one set of muscles changes state with respect to an increase of tension, an antagonistic set may become relaxed in a reciprocal exchange of entropy of configurational free energy. Thus changes of state may occur in sets of flexor and extensor muscles, as the body as a whole remains in a state of constant free energy. Thus it is possible to define such a resting state of basal metabolism as the most accessible state. This state is maintained by phase rule conditions of invariance and constraint, as applied to the body as a whole (*Joseph,* 1971a, b; 1973).

The second law of thermodynamics (*Carnot*'s principle) has been put into the form of conservative line integrals, which yield null values for all reversible processes in invariant isothermal systems (*Caratheodory,* 1909; *Born,* 1948). In this form, the second law may be stated, 'there are inaccessible states in the neighborhood of any given state'. This principle explains the great stability of states of maximal stability in the mammalian organism. Thus, according to *Carnot*'s principle, the values for the line integrals for all reversible processes of transport of water or ions approach zero in an invariant biological system. Such a conservative system remains always in this accessible state. Except for redistributions of energy, entropy and muscular tone or tension within the organism, the body remains in a standard homeostatic state of energy and electrolyte balance. In this state, there are redistributions of free energy and tension, but under phase rule conditions of invariance and constraint that maintain the total energy and entropy at constant values.

When the body goes into a state of production of external work, the behavioral response must conform to modified conditions of constraint. The muscles change state to secondary states of normal accessibility, determined by the nature of the physical activity of walking, running, or producing physical work. The active state is then one of modified accessibility; it is characterized by a set of physicochemical parameters that depend on the nature of the behavioral responses. Thus if the resting accessible state may be characterized by the set of parameters:

$$S = (S_1, S_2 \cdots S_n),$$

these may be taken to refer to the cellular structures $C_1, C_2 \cdots C_n$.

In a well-ordered response that involves only one degree of freedom,

$$S \rightarrow S'$$

where

$$S' = (S_1', S_2' \dots S_n')$$

Here S' denotes an accessible state of n cellular parameters which are well-ordered but in a modified state of balance with respect to energy, electrolytes and water. This response occurs with one degree of freedom when both sets of physiological and physicochemical parameters are well-ordered, and when they correspond to well-defined resting and active states. These considerations follow from the application of *Carnot*'s principle as developed in *Caratheodory*'s derivations relating to well-ordered cyclic reversible processes.

Chapter 2

Logical Principles

In examining the logical basis of a theory of biological behavior, it is necessary to presume the preexistence of certain sets of judgements. In general, these are of the nature of value judgements that differ in kind from those made in the nonbiological sciences. It is generally assumed in physics and chemistry that the objects of our studies, inert objects, forms of energy in relation to material substances or interactions of atoms and molecules are indifferent to animal feeling, sensations or perceptions. Their responses are presumed to be *non-agapastic*. This judgement cannot be presumed in the study of animal or human behavior, even in the anesthetized subject. Living cells and tissues always exhibit the properties of irritability.[1] Any change of condition always acts as a stimulus or as an inhibitor; the precise response depends on conditions of constraint over which the observer may have little or no control. Thus biological experimentation generally lacks the complete objectivity of physical or chemical experimentation, in which a few primary conditions can be brought under exact constraints. In biological or behavioral studies, the results often depend on numerous unobservable conditions of constraint that cannot be exactly reproduced.

It is also true that the conditions under which a field naturalist or ecologist such as *Wheeler* makes his observations are completely different from those that apply to the studies of an experimentalist such as *Pavlov*. The conclusions drawn by any such biologist depend on value judgements that are presumed in his

---

[1] This is a fundamental preconception dating from the seventeenth century, when the concept of inherent irritability was introduced by *Glisson* (1650). It has been accepted as a first principle for at least three centuries. Nevertheless it has recently been stated by *Lew* (1978) that 'the maintenance of the large Ca ion gradient across the plasma membrane of living cells is an essential condition for cell survival and for the various physiological functions of calcium'.

An alternative formulation of the problem has however been given (*Joseph,* 1966, 1973). According to this, the existence and survival of the living cell is the *primary* function, and all other functions including metabolism, are *secondary*. The primary properties survival, irritability and behavior, are *existents* or *essences,* which cannot be inferred from the secondary properties that depend on morphology, phylogeny, ontogeny, and genetics.

fundamental preconceptions. They also depend on judgements based on the work of his predecessors. In every case the conclusions drawn from the work of biologists in any field require criticisms of the fundamental value judgements that have been applied to their fundamental conceptions and presuppositions. These presumptions are both positive and negative – the preconceptions that are included as well as those that are omitted.

In general, there are great differences in the fundamental preconceptions made by scientists – even by contemporaries in the same field of study. For example, biologists differ greatly in their approaches to questions of organism and mechanism. Following the Cartesian approach are the reductionist physiologists, who emphasize the study of one specialized function: nerve conduction, endocrine secretion, nutrition and so forth. *Pavlov* and an entire school of physiologists who tend to derive behavior from systems of nervous reflexes are primarily reductionists. On the other hand, *Sherrington,* who has employed reductionist methods most effectively in all his work, has also adopted organicist conceptions in *The Integrative Action of the Nervous System* (1920), and in *Man on His Nature* (1940). *Cannon*'s *The Wisdom of the Body* (1929) is organicist in its approach. In this he continues the tradition established in experimental medicine by *Bernard* (1865, 1878, 1927).

In contrast to the organicists, large groups of influential physiologists have apparently been dominated by mechanist and reductionist views. These include students of isolated nerves, muscle fibers or neuromuscular preparations, which attempt to isolate the mechanistic basis of nerve conduction, ion transport, electrical transport and membrane permeabilities apart from the integrative or emergent basis of the nervous system as a whole, including the brain and the sense organs. The fundamental preconception here seems to be that behavior can be studied independently of the emergent properties of the submicroscopic morphology of protoplasm. This has led to the multiplication of hypothetical entities such as ion pumps, energy transport systems, and responses of membranes, viewed as independent entities rather than as emergent wholes.

Among biochemists, the great majority have been trained to accept the Cartesian or reductionist approach. This consists in isolating or attempting to isolate enzymes, coenzymes and intermediary metabolites, assigning to each substance an individual function in the total metabolism. Often the organistic viewpoint has been subordinated to the independent behavior of the individual substance. At present, these are numbered in the hundreds or thousands, each of which contributes in some way to the metabolism of carbohydrates, lipids and proteins in an elaborate chain of cause and effect relationships.

The predominant function of intracellular water, which constitutes by far the bulk of the cellular mass, seems to have been almost forgotten. In its various states of aggregation, water determines the dielectric properties, solubilities and all the other thermodynamic functions and reaction rates of the system (*Joseph,*

1973). It thus serves to unify and coordinate all the functionsof the cell as a whole, and thus to behave as an emergent property of all the units of the chemical morphology. This organismic view of the properties of protoplasm is supported by the views of *Bernard, Frey-Wyssling, Willstatter* and *Oparin,* cited in chapter 1. The properties of water in biological systems are colligative, and depend on the physicochemical state of entire sets and aggregates. Therefore, the emergent properties can be considered only from the principles of *Carnot* and of *Gibbs* rather than from mechanistic principles, which are fundamentally Cartesian, dualistic and reductionist.

Emergent Properties of Thermodynamic Functions

A *physicochemical system* in biology may be defined as an aggregate of several components and phases. In the definition of *phase* by *Gibbs* (1875, 1928), 'We may call such bodies as differ in composition and thermodynamic state different *phases* of the matter considered, regarding all bodies which differ only in quantity and form as different examples of the same phase.' Living mammals, from the point of view of physical chemistry, are therefore heterogeneous systems consisting of colloidal components, water and electrolytes distributed among a large number of intracellular and extracellular phases; only a limited number and kind of these phases can be regarded as homogeneous solutions. These include body fluids such as lymph, peritoneal fluid, synovial fluid, aqueous humor, and other liquid phases of the milieu intérieur. These have the approximate water and electrolyte composition of dialysates of blood plasma, which represents the common properties of the milieu intérieur.

The fluids have in common water as the dispersion medium and, in addition, four electrolyte components: $NaCl$, KCl, $CaCl_2$, and $MgCl_2$. Other electrolytes including phosphate, sulfate and bicarbonate ions are present at low concentrations in equilibrium with hydrogen ions. The chemical potential of water, μ_{H_2O}, and the freezing point depression, Δ, are colligative properties related in the following way:

$$\mu_{H_2O} = \mu_{H_2O}{}^0 + 5.26\ \Delta \tag{2.1}$$

where $\mu_{H_2O}{}^0$ is the chemical potential of pure water at the freezing point, and Δ is the freezing point of the fluid in centigrade degrees. The constant 5.26 is the entropy of fusion of ice, measured in calories per mole per degree. The chemical potential of water in blood plasma is of the order of about -3.1 cal/mole, and is nearly independent of age and species for all mammals. It is thus a phylogenetic and ontogenetic invariant.

According to the phase rule, μ_{H_2O} is also constant in all phases of the heterogeneous system that are in equilibrium with blood plasma. It is thus subject to thermodynamic conditions of constraint that depend on the standard chemical morphology of each kind of solid, semi-solid or liquid phase. The fundamental condition of constraint is given by *Gibbs'* equation 77:

$$\mu_{H_2O}' = \mu_{H_2O}'' \tag{2.2}$$

This depends on the reversible transport of water between any two phases of the heterogeneous system (*Gibbs*, 1875; 1928; *Joseph,* 1971a, b; 1973). Similar conditions of constraint apply to all the inorganic ions (Na, K, Ca, Mg, Cl) that are transported reversibly between any two phases. The conditions for reversible transport are given by *Gibbs'* equation 77, which yields:

$$\mu_{AB}' = \mu_{AB}'' \tag{2.3a}$$

and

$$\mu_{AB_2}' = \mu_{AB_2}'' \tag{2.3b}$$

where AB refers to a binary electrolyte of the type NaCl or KCl, and where AB_2 refers to ternary electrolytes such as $CaCl_2$ and $MgCl_2$.

Equations 2.2 and 2.3 are the conditions for reversible isothermal transport of water and ions in a heterogeneous biological system. Since the chemical potentials remain invariant over long periods of time, they are conservative properties that depend on reversible transport and exchange. Their invariance depends on a principle which may be called 'the conservation of reversibility' (*Joseph,* 1971, 1973). The principle is derived from the second law of thermodynamics (*Carnot*'s principle), which implies that the free energy change of reversible processes in an invariant system is zero. This is expressed by the law:

$$\Delta G_{rev} = -W_{max} = 0 \tag{2.4}$$

where ΔG_{rev} is the free energy change, and W_{max} is the maximal work of the reversible process. The condition for stability of composition in an invariant biological system implies a null value for the reversible work. This implies constant values for the chemical potentials of water and the inorganic electrolytes.

According to *Gibbs'* equation 97, these chemical potentials in any phase of the system are mutually related at constant temperature and pressure. Thus:

$$-m_{H_2O} \, d\mu_{H_2O} = \Sigma \, m_i \, d\mu_i \tag{2.5}$$

where m_{H_2O} and m_i refer respectively to the mass of water and each kind of ion expressed as number of moles. Thus the chemical potential of water is an emergent property of the composition of each kind of phase. Since this remains constant in all phases, electrolyte compositions of the phases are never independent, but represent emergent properties of the entire organism. They then behave as emergent sets or aggregates with common properties:

$$S' = (S_1', S_2' \ldots S_n')$$

and

$$S'' = (S_1'', S_2'' \ldots S_n'')$$

where

$$S_1' \supset S_1''$$

and where the symbol $\supset$ is read as 'corresponds to'.

Thus all sets of thermodynamic properties that depend on chemical morphology remain constant when the chemical morphology of all phases remains invariant. It differs, however, for each kind of phase, depending on the state of aggregation and the dielectric properties of water as a dispersion medium. Each member of one of the sets, S', is in biunivocal correspondence with its correlate in any other set, S'', of the invariant system. When the morphology changes in time, as in growth, development and aging, each set is an emergent property of the organism as a whole, and is subject to thermodynamic conditions of constraint. This is implied by the relation:

$$S = (M, N, O, P \ldots)$$

where S is the set of all phases in the organism. At any time, M, N, O, P ... are related to each other by biunivocal correspondences which depend on the state of the organism as a whole.

In an invariant physicochemical system, the free energy, G, is an invariant emergent property that depends on the invariant potentials of water and electrolytes. Thus:

$$G = \Sigma \, m_i \, \mu_1$$

where the summation has taken over all the chemical potentials that are functions of state.

When the milieu intérieur is invariant in time, every intracellular phase remains invariant with respect to μ_{H_2O}, μ_{AB} and μ_{AB_2}. However, in binary and ternary electrolytes,

$$\mu_{AB} = \mu_A + \mu_B$$

and

$$\mu_{AB_2} = \mu_A + 2\mu_B$$

Therefore the constraints do not imply constant ontogenetic values of μ_A and μ_B but restrict them only by conditions of the form

$$\Delta\mu_A + \Delta\mu_B = 0$$

for binary electrolytes, and

$$\Delta\mu_A + 2\,\Delta\mu_B = 0$$

for ternary electrolytes. In every case, $\Delta\mu_A$ and $\Delta\mu_B$ depend on intracellular states of aggregation which may vary in growth, development and aging. Thus intracellular water and electrolyte composition may vary in time, but only under phase rule conditions of invariance.

Therefore, all the intracellular ion potentials, $(\mu_A, \mu_B \ldots \mu_i)$, may be functions of time under conditions in which $\mu_{H_2O}, \mu_{AB}, \mu_{AB_2}$, etc. remain invariant. Hence, the thermodynamic properties have the character of emergents of chemical morphology as it develops in time. However, all the emergent thermodynamic functions are implied by the emergent properties of the organism as a whole. These properties include physiological behavior, which is also constrained and restricted by chemical morphology and physicochemical state, as they develop according to biological laws of growth and development. Since behavior depends on changes of state of the well-ordered submicroscopic phases of protoplasm, it may be regarded as a colligative property of all intracellular and extracellular phases. The colligative properties of any phase are those that depend on the chemical potential of water, as it is determined by the composition and state of the other components, especially the inorganic physiological ions. Muscular tension, for example, depends on the dielectric energy, as related to the standard chemical potential of the sodium ion (*Joseph,* 1971a, b; 1973). This in turn depends on the state of aggregation of water, which determines its dielectric properties, as related to tension and length. Many such parameters related to neuromuscular behavior are colligative properties, which depend on the physicochemical state of intracellular aggregates. Changes of these properties depend on changes of state of aggregation which occur in biological processes related to behavior.

Mathematical Inference

'The point of mathematics is that in it we have always got rid of the particular instance, and even of any particular sorts of entities' (*Whitehead,* 1925). Inferences based on the formulations of pure mathematics are therefore of the broadest generality; they lead to quantitative predictions and meet the strictest criteria of objective verification. Compared to mathematical inference, the generalizations derived from traditional formal logic (laws of thought) are not subject to quantitative predictability and verification. They often depend on inferences from special cases (hypotheses) or on questionable generalizations based on inadequate sampling (induction). Attempts to verify hypothetical formulations often lead to a multiplication of ad hoc entities or mechanisms and to a violation of *Occam*'s razor – 'entities are not to be multiplied without necessity'.

Mathematical generalizations are of no value unless they are based on universal laws, rather than on special limiting cases. Thus, attempts to derive electrolyte distribution from classical theories of membrane equilibrium and membrane potentials are of no value if the thermodynamic principles on which such theories are based represent only limiting cases rather than the actual generalized laws of heterogeneous equilibrium. It has been shown in earlier studies (*Joseph,* 1971a, b; 1973) that the most valid generalities are to be found in *Gibbs'* (1875, 1928) *Equilibrium of Heterogeneous Substances* rather than in *Donnan*'s (1911) *Theory of Membrane Equilibrium and Membrane Potentials in the Presence of a Non-Dialyzable Electrolyte.* As the title shows, the '*Donnan* equilibrium' is based on the coexistence of two miscible solutions or phases separated by a semi-permeable dialyzing membrane. Each solution is a homogeneous system characterized, like all such systems, by perfect mixing.

The validity of *Donnan*'s formulations depends on the verifiable truth of this presupposition. *Gibbs'* theory, as is shown by his title, is based on the general thermodynamics of *immiscible* heterogeneous phases; it can lead to *Donnan*'s special formulation only in the limiting case of a homogeneous solution. According to this formulation, the structure of heterogeneous biological systems depends on the special properties of the *membrane* in each case, rather than on the well-ordered submicroscopic structure of protoplasm. Therefore, the theory of 'membrane equilibrium' is based not on the characteristic chemical morphology of protoplasm in a given state of aggregation, but rather on the unobservable properties of ion pumps, selective permeabilities, and irreversible metabolic processes. The result has been that all subsequent developments of the theory of active transport have been based on nonverifiable inferences regarding the nature of the cell membrane. This has yielded a theory which takes no account of submicroscopic morphological structure or the properties of water as an organized part of protoplasm, with altered dielectric and solvent properties. It also takes no account of the functions of water in relation to *changes* of metabolism or to changes of structure in behavioral responses. These fallacies are results of the failure to base a theory of chemical morphology and physicochemical state on the most general thermodynamic theory of heterogeneous structures. As a result of this failure, the mathematical formulations do not represent the general laws based on *Carnot*'s principle and the phase rule. Hence the derived formulae do not 'get rid of the particular instance and even of any particular sorts of entities'. In the present form of 'active transport', these special entities are multiplied without necessity and without any apparent limit. Hence *Whitehead*'s 'point of mathematics' is violated, because the theory depends on all sorts of 'particular entities'.

Valid mathematical inference depends on finding its basis in the general laws of nature rather than in special cases, as in the '*Donnan* equilibrium'. As *von Helmholtz* (1847) has remarked, 'The principle of causality is nothing else but the supposition that all the phenomena of nature are subject to law.' Subsequent development of thermodynamics has led to *Gibbs' Equilibrium of Heterogeneous Substances,* based on *Carnot*'s principle. The second law, as has been previously pointed out, has been put in its most general mathematical form by *Caratheodory* (1909). The general principle can be stated in the form 'there are inaccessible states in the neighborhood of any given state'. According to this, protoplasm normally exists in an accessible state.

Carnot's principle prevents a redistribution of ions that would lead to a state of high intracellular sodium and low intracellular potassium. This never occurs in human skeletal muscle or heart. Such a redistribution would represent an inaccessible state forbidden by *Carnot*'s principle. Actual physicochemical state is normally stable, reversible and invariant, and independent of irreversible mechanisms (*Joseph,* 1971a, b; 1973). All physicochemical and physiological properties are therefore necessary properties of the primary function, which is the existent chemical morphology. Like all secondary thermodynamic functions, these are emergent or colligative properties that depend on chemical morphology as related to the possible states of aggregation of water and to its dielectric or solvent properties. The dielectric properties are factors in cellular metabolism, which therefore changes with states of aggregation.

Logical Inferences

Logical inference is based on classification rather than on strict mathematical formulations. Classification is a necessary half-way step toward strict mathematical formulation. However, it could be concluded that even a partial formulation based on *correct* classification of any problem is preferable to elaborate formulations based on incorrect or half-true mathematical preconceptions or suppositions. The necessary basis of any theory of biological structure or behavior requires its classification as a physicochemical system. Testing of logical inferences depends on the study of certain typical forms known as *syllogisms.* This is true of inferences based on simple subject-predicate relations that involve what is known as a 'middle term'. When the relations involve subject or predicate as emergents of entire sets or classes of individual entities, it may not be possible to apply simple syllogistic or cause-effect logic.

A fundamental syllogism of formal 'subject-predicate' logic is of the form: S is M; M is P; therefore S is P.

Here S stands for subject, M represents the middle term, and P is the predicate. The syllogism is analytical, deductive and a universally valid inference if the two premises are based on valid judgements. It may be put into the form:

Rule: Subject is middle term.

Case: Middle term is predicate.

Result: Subject is predicate.

An example of an analytical inference in physiology might be:

Rule: All muscles can contract.

Case: M is a muscle.

Result: M can contract.

An inversion would yield:

Rule: All muscles can contract.

Result: M can contract.

Case: Therefore M is a muscle.

This is an example of *hypothesis,* a form of synthetic inference. The hypothesis may or may not be true. An experimental test might show M to be a rubber band rather than a specimen of muscle.

Another inversion might be of the form:

Case: M is a muscle.

Result: M can contract.

Rule: All muscles can contract.

This is an *inductive* inference. It might be valid for all muscles under physiological conditions of specified temperature and electrolyte composition of the surrounding salt solution. In general, the validity of the inference depends on the conditions of constraint in any set of experimental conditions. Under well-controlled conditions that permit contraction in *any* case, the induction would be valid. It would not be valid in solutions that contain certain inhibitors, or that do not contain the correct concentrations of physiological ions. Therefore, inductions must always be subjected to experimental test. Similarly, the validity of the analytical or deductive syllogism depends on the truth or validitiy of the major premise. This is a matter of judgement. To be perfectly valid, the major premise might be required to state: all muscles can contract under the proper conditions of constraint.

The syllogisms of formal logic are therefore generally required to be properly qualified with respect to their major and minor premises — these must be valid as true or false *judgements.* If either premise is untrue, the inference is not generally valid.

Examples of fallacious syllogisms from current work in cellular physiology may be cited.:

Major premise: Electrolyte distribution (composition) depends on temperature.

Minor premise (unstated): Temperature depends on metabolism.

Conclusion: Therefore electrolyte distribution (composition) depends on metabolism.

At present, this false conclusion has been widely accepted as a major premise or fundamental preconception of the theory of 'active transport'. It depends on the unstated or implied minor premise that *temperature* depends on metabolism. This implies that metabolism is the primary property, and that temperature is secondary and dependent on metabolism. But in biology the minor premise is clearly false and a faulty judgement. In biological organisms such as cold-blooded vertebrates or invertebrates, metabolism (M) is secondary, and temperature (T) is primary. Hence it is true that

$$T \supset M$$

and false that

$$M \supset T$$

where the symbol $\supset$ is read as 'implies'. Fish and invertebrates come to the approximate temperature of the ocean, lake or river, and preparations of cells and tissues come to temperature equilibrium with refrigerators or cold rooms. Consequently the unstated minor premise of the above syllogism is incorrect as a hypothesis, and the conclusion that electrolyte distribution depends on metabolism (active transport) is unacceptable.. This is an example of a general kind of fallacy that occurs in formal logic — omission of the middle term. A correct formulation of the argument requires the valid judgement that

$$T \supset (M, \text{composition}).$$

Thus temperature implies simultaneous changes of metabolism and composition (chapter 3). Then all intracellular changes of reaction rates and physico-chemical state are functions of a *process* (P). In this case the process is that of placing the preparation of cells or tissues into a cold environment. Then:

$$P \supset (T, M, \text{composition}).$$

Neither metabolism (M) nor composition can be regarded as simple terms in an Aristotelian syllogism, or as simple functions of temperature or process. They must be regarded as univariant sets, which behave with one degree of freedom with respect to an independent variable or argument. With respect to changes of temperature, the system has one degree of freedom. In the above experiment, both respiratory metabolism and chemical morphology are functions of one independent variable, the temperature. To omit this as the middle term of a syllogism is to commit an error of judgement, by falsifying a natural condition of constraint in biological organisms. Temperature is a primary function, and respiratory metabolism is secondary. Respiration is not the independent variable. It depends on physicochemical state, and is a function of temperature and of intracellular chemical morphology and physicochemical state. To find a conceptual scheme which permits cells and tissues to change with one degree of freedom in responses to any change of external conditions, it is necessary to find the most valid universe of logical discourse. This cannot be mechanistic in nature, but must depend on the emergent properties of organismic physicochemical systems.

The Universe of Discourse

Significant discussion of the relationships between physicochemical, physiological, metabolic and behavioral properties of biological structures on one hand and morphology and species on the other depends on establishing the conditions for a logical universe of discourse. 'The classification of any biological system as a physicochemical system is the first step in finding the laws of its structure and behavior' (*Joseph,* 1973). This requires a classification of cells and tissues as complex coherent systems that satisfy a number of necessary and fundamental conditions.

Every living animal at the level of vie constante (mammals and birds) satisfies the following set of conditions in the adult state. It is a member of a definite warm-blooded species that is characterized by a certain number of invariant properties common to all normal members of the species, independently of age and sex. It must therefore meet certain conditions of invariance with respect to chemical morphology, constant standard rates of basal metabolism and a standard state of heterogeneous morphology — including the normal set of histological, histochemical and anatomical relationships. These complex coherent properties then imply the coexistence of three fundamental conditions:

(I) Day to day invariance of chemical composition of the body fluids and of intracellular phases.

(II) Constant chemical reaction rates in homeostatic steady states.

(III) Normal invariant heterogeneous structure of all cells and tissues.

The physicochemical and metabolic states of all members of the species must satisfy the set of three coherent conditions. This implies that all members of the species belong to a common universe of discourse, IV, characterized by a set of properties common to I, II and III. This common universe of discourse implies an organism that is invariant, reactive and heterogeneous. Unless this criterion is fulfilled, the physicochemical state does not represent the conditions for a normal living animal.

In the theory of 'active transport', the conditions are those of a homogeneous intracellular phase contained within a semi-permeable membrane. An invariant state of water and electrolyte balance is sustained, not by phase rule conditions of constraint, but by mechanistic processes that require metabolic sources of energy. By certain investigators, water balance is maintained by 'water pumps', likewise subject to mechanistic control. The universe of discourse does not require thermodynamic conditions of constraint that maintain invariant heterogeneous physicochemical states that imply constant irreversible metabolic rates and transport of oxygen, carbon dioxide and heat. It therefore requires ad hoc mechanisms that maintain thermodynamically unstable mixtures of water and electrolytes. It has been argued that these arbitrary mechanisms are inferences that result from an indadequacy of the universe of discourse to which the results are referred. This is a deficiency or inadequacy of the conceptual scheme that is presupposed. An intracellular homogeneous solution cannot at the same time fulfill the conditions of reactivity and invariance. It is unable to control metabolism and electrochemical properties by a univariant change of state of water, which implies simultaneous changes of reactivity, metabolism and all other secondary properties by a univariant change of state. This is implied by the behavior of univariant sets of the kind

$$A, B, C \ldots \supset X, Y, Z \ldots$$

where X, Y, Z ... denotes a set of secondary properties that changes simultaneously with a primary set A, B, C ... When the change of state can be represented by one independent variable (the dielectric constant D'', for example), the process

$$D'' \rightarrow (D)'$$

implies a change of all the secondary properties

$$(X, Y, Z \ldots) \rightarrow (X', Y', Z' \ldots).$$

Such changes of state are univariant and well-ordered. They imply well-ordered behavioral responses to an intracellular change of state resulting in a univariant response of the state of water or the dielectric properties. These are colligative properties that depend on changes of chemical potential of all the inorganic ions. Thus complex behavioral responses involve simultaneous changes of chemical potentials and electrical potentials, behaving as secondary properties as referred to morphology. The responses also involve changes of intracellular respiratory metabolism, which also reacts as a secondary property.

Causality

The number of neurones in the human brain is said to be of the order of 10 billion, and the total number of all the cells of the neuromuscular system must be a large multiple of that figure. At any time human behavior would be reciprocally reflected in physicochemical changes of state in all cells and tissues. To express this as a cause-effect relationship between morphology and behavior in every detail would be impossible.

An alternative approach would be to regard the relation beween organismic and physicochemical behavior to be of the nature of a reciprocal mirror image of all the microcosmic processes in the body. Conversely, the microcosmic processes reflect the sum total of external relationships between organism and environment. In this way we get rid of detailed cause-effect relationship in a way that is something like *Leibniz'* solution of the problem. According to him, the ultimate living units are 'windowless monads' which behave according to a preestablished harmony governed by a principle of sufficient reason. This is the solution to the problem of the 'labyrinth' in which we find ourselves entrapped whenever we seek to disentangle existential problems by the method of seeking ultimate causation.

The question of solving scientific problems by means of ultimate cause-effect relationships has had a long and tortuous history. According to *Newton,* the solution of general problems of cosmology were approached by means of universal laws. Particular or special problems were studied in the light of immediate causality. *Kant* and many of his contemporaries considered every phenomenon in the universe to have an immediate cause. At that time causality was the ultimate scientific explanation. Two of the foremost physicists of the 19th century, *von Helmholtz* and *Mayer,* expressed themselves as follows: '... causality is nothing else but the supposition that all the phenomena of nature are subject to law' (*von Helmholtz,* 1847). According to *Mayer* (1842): 'Forces are causes, and consequently it is reasonable to apply to them a principle *"Causa aequat effectum".'*

Peirce (1892) tends to agree with these views. 'Whether we ought to say that a force *is* an acceleration or that it *causes* an acceleration is a mere question of propriety of language.' It might be said that *Peirce* agrees with *von Helmholtz* and *Mayer* in regarding effects as identical with causes. This tends to reduce all causality in nature to laws of identity and to eliminate elaborate causal explanations in biology. This tendency has been strengthened in the 20th century by the

organistic views of *Whitehead* (1929). Here mechanistic views have been practically eliminated in a philosophy which regards natural processes as the ultimate
reality, to which mechanistic causality at lower emergent levels is subordinated.

It is a natural extension of this line of thought that external and internal
behavior in man and other organisms are natural processes which are related by
one-to-one biunivocal or mirror image projections. To go a step further, it would
be possible to say that the internal mirror image takes the form of multitudinous
changes of state of water in different states of aggregation. These are colligative
properties that are related to multitudinous secondary processes that involve
metabolism, tensions, irritabilities, muscular contraction, nerve conduction and
numerous related secondary phenomena.

Synechism

The following definitions of the concept of synechism were written by
Peirce for *Baldwin's Dictionary of Philosophy and Psychology* (1902).

Synechism (Gr. συνεχισμός) continuous holding together; not in use in
other languages. That tendency of philosophical thought that insists upon the
idea of continuity as of prime importance in philosophy, and in particular, upon
the necessity of hypotheses involving true continuity.

'A true *continuum* is something whose possibilities of determination no
multitude of individuals can exhaust. Thus, no collection of points placed upon
a truly continuous line can fill the line so as to leave no room for others,
although that collection had a point for every value towards which numbers
endlessly continued into the decimal places would approximate, nor if it contained a point for every possible permutation of all such values. It would be in
the general spirit of synechism to hold that time ought to be supposed truly
continuous in that sense ... To suppose a thing inexplicable is not only to fail to
explain it, and so to make an unjustifiable hypothesis, but much worse — it is to
set up a barrier across the road of science, and to forbid all attempts to
understand the phenomenon.'

In preceding sections of this chapter, the hypothesis has been made that
biological organisms, as individuals and as members of communities or associations in a complex *biocoenose,* are coherent aggregates. This fits the definition
of synechism, which means a continuous holding together over long periods of
biological time. Thus the life of any animal or plant is a synechistic whole. It is
continuous and unified over the entire life span of any individual. Moreover, it is
continuous with the life of the species, which has persisted continuously (or
synechistically) for thousands of generations.

The hypothesis of continuity is necessary to account for morphological
invariance in any species: this is a fundamental preconception necessary for the

concepts of speciation and taxonomic classification for all the phyla of plants and animals. These concepts were accepted in the ancient world by *Aristotle,* and by all modern biologists since the time of *Linnaeus.*

Thus the concept of synechism is of fundamental importance in biology to yield the general principle of well-ordered hierarchical sets in the general universe of discourse which assumes living organisms to be well-ordered, complex and coherent physicochemical aggregates. The further hypothesis that animal behavior is a function that is 'held together' and continuous with species and morphology also depends on a principle of continuity. Protoplasm and its secondary physicochemical and physiological properties are coherent and synechistic. Biological laws must take account of this continuity and coherence. The further hypothesis that animal behavior is a function that is 'held together' and continuous with species and morphology also depends on a principle of continuity. Protoplasm and all its properties are coherent and synechistic in time. Biological laws must take account of these characters.

Tychism

Peirce's definition of *tychism* is also to be found in *Baldwin*'s dictionary (1902). It is from the Greek word τυχισμός, chance, not is use in other languages.

'The mere proposition that absolute chance is operative in the cosmos may receive the name of tychism.' The term appears to be first used as follows: 'I endeavored to show what ideas ought to form the warp of a system of philosophy, and particularly emphasized that of pure chance ... I argued further in favor of that way of thinking which it will be convenient to christen *tychism.*'

The doctrine of chance or probability is essentially mathematical. Antecedents are found in 17th century mathematics, especially in problems formulated by *Pascal* and by *Fermat. Peirce*'s concepts of tychism and synechism tend to give chance and necessity (freedom and necessity) equal shares in the process of evolution. In the history of philosophy, the balance has tended to shift between the two extremes depending on the climate of opinion in any given period.

Due to 20th century developments in physics, opinion has moved strongly toward the direction of absolute chance. This has been a result of developments such as *Heisenberg*'s principle of indeterminacy. The tendency has been to base the laws of physics on what is essentially a statistical foundation. For elementary particles at the atomic or subatomic levels, individual events are, to a great extent, arbitrary and indeterminate. Large-scale phenomena are macroscopic, emergent and to a great extent follow the principle of necessity, but not absolutely. As the dimensional scale is reduced, phenomena such as Brownian movement appear, and these are statistical or tychistic.

In the latter part of the 19th century, the theory of gases and other states of matter was based on kinetic and statistical principles, or on the laws of chance. The science of statistical mechanics was developed by *Boltzmann, Maxwell, Gibbs* and *Einstein.* This made it possible on a statistical basis to develop theories of Brownian movement and of the thermal properties of molecular aggregates. Thus the realms of chance and necessity tend to converge on the basis of statistical laws. In chapter 3, it will be shown that the laws of thermodynamics may be related to mathematical principles of probability or pure chance. Fluctuations due to chance are also operative in Brownian movement, reaction rates, spreading and diffusion. These are universal large-scale phenomena that are always operative in animal behavior and evolution. They may become dominant in lower forms of life (vie oscillante), which according to *Bernard* (1878) have not evolved complete freedom from chance events in the environment. Higher organisms (vie constante) have, however, acquired a high degree of independence from random external events. This has made possible the development of internal freedom from external necessity governed by chance.

Chapter 3

Entropy and Probability

A crystalline solid is a well-ordered state of matter. Many chemical substances which are liquid or gaseous at room temperature become crystalline solids at very low temperatures. These include oxygen, nitrogen, carbon dioxide, and even hydrogen, which is most difficult to liquefy. At absolute zero the entropies of many pure crystalline substances become zero; for others the entropy approaches zero. In this state the spatial coordinates of the atoms and molecules are perfectly fixed. Along with the kinetic energy, all the disorder and randomness have been taken out of the system. Temperature is associated with the thermal kinetic energy; when this is gone, all that remains is the chemical bond energy. The electrons retain their quantized orbital energies, but all the atoms are in their unexcited ground states.

Perfect order exists in a molecular kinetic assembly only at the absolute zero of temperature. This is a highly improbable state of matter, and one that is very difficult to achieve in the laboratory. All other states of matter are comparatively chaotic; the molecules have kinetic energy which is distributed among translational, rotational and vibrational motions. At all temperatures energy is distributed according to the laws of statistical probability. The more ways in which the energy can be distributed and the more ways in which the molecules may be arranged in space, the higher is the entropy. Thus the molal entropy of a gas mixture is higher than the sum of the entropies of the gases. The permutations and combinations of molecular energies and spatial coordinates increase as energy is added to the system; they increase also with the numbers of molecular species which are present. An increase of the number of components causes an increase of the entropy. The way in which entropy is related to the distribution of molecular energies is studied by the methods of statistical mechanics.

The physical laws which derive from the energy and entropy principles apply to large assemblies of atoms and molecules. Solubilities and vapor pressures, for example, are bulk properties; the concepts are not applicable to extremely small quantities. The assemblies that are observed in the laboratory are usually sufficiently large to conform to statistical laws. 18 g of water vapor contain 6.02×10^{23} molecules; 18 μg contain 6.02×10^{17} molecules. These are within the usual scale of observable properties. Assemblies of a few hundred or a few thousand molecules in the liquid or solid state are difficult or impossible to study. The instruments of measurement are designed for studies on a larger scale. Under the circumstances, measurements of the bulk properties such as mass, temperature, volume or pressure, become all but meaningless. Energy and the related thermodynamic functions apply to phenomena which are not necessarily large scale, but which are well beyond the level of ultramicroscopic assemblies of atoms and molecules. The laws of statistical probability are the laws of large numbers.

Probability

Everyone knows what is meant by probability, but no one can define it exactly. It is what *Russell* has called a 'heterologous word'. The word 'short' is short, but the word 'long' is not long. 'Short' is autologous: it describes itself. 'Probability' at the lower limit means 'impossibility'; at the upper limit is means 'certainty'. Midway there is something called the 'even chance'. A future event is probable when there is more than an even chance that it will occur; it is a matter of speculation. It is no accident that the mathematical theory of probability is related historically to the theory of games of chance.[1]

In a bridge deck there are 26 black and 26 red cards. The probability of drawing a heart from the full deck is one chance in four. The probability of drawing two consecutive hearts is:

$$\frac{1}{4} \times \frac{12}{51} = \frac{1}{17}.$$

This assumes perfect mixing and the equivalent of blindfold selection. For an extremely large deck (52,000 cards containing 13,000 hearts), the probability of drawing a second consecutive heart is almost exactly equal to the probability of drawing the first one. Three rules are applicable:

(1) The probability, p, of any event is the ratio of the number of possibilities that it will occur to the total number of possibilities.

(2) The probability of any successive event is the ratio of the number of remaining possibilities that it will occur to the total number of remaining possibilities.

---

[1] The mathematical theory of probability seems to have originated about 1660 in the minds of *Pascal* and *Fermat* (perhaps independently). In a paper submitted to the Academie parisienne des sciences, *Pascal* proposed, among other projects, to reduce to an exact art, with the rigor of mathematical demonstration, the incertitude of chance, thus creating a new science which could justly claim the stupefying title: the geometry of hazard.

(3) The probability of two successive events is given by the product of the two individual probabilities.

The probability of a series of events is then:

$$P = p_1 \times p_2 \dots \times p_n$$

For example, the probability of drawing 13 successive hearts from a deck of 52 cards is

$$P = \frac{13 \times 12 \times 11 \times 10 \dots \times 1}{52 \times 51 \times 50 \times 49 \dots \times 39}$$

The product in the numerator is called 'factorial 13' and is denoted as 13! The probability of the first card being a heart is one in four. It is impossible to draw 14 consecutive hearts.

It has been stated that entropy is related to probability. We know that the entropy of oxygen and nitrogen increases as the two gases are mixed at constant total pressure. The entropy would decrease if the two gases could be separated. The possibility of decreasing the entropy depends on the number of molecules in the mixture. To state the problem in terms of probability, a macroscopic case may be visualized. Suppose that we have 2,000 beans consisting of 1,000 black and 1,000 white. They are thoroughly shaken and mixed, forming a nearly homogeneous mixture. The beans are then sorted mechanically into two compartments, the operation being one of pure chance. Half the beans are destined to fall into compartment A, the other half into B. The probability for each white bean to fall into A is assumed to be exactly $\frac{1}{2}$. The probability that all of them fall into A is then

$$p = \left(\frac{1}{2}\right)^{1,000}$$

or about 10^{-300}. As one would expect, this is extremely low; however, it is not an absolute impossibility. If the number is taken as $N = 6.02 \times 10^{23}$ (*Avogadro*'s number) P would represent the probability of perfect separation of $\frac{1}{2}$ mole of nitrogen from $\frac{1}{2}$ mole of oxygen in a mixture of equal numbers of the two kinds of molecules. The gas mixture represents an almost infinitely more probable distribution of molecules than the pure gases. The mixture is in statistical equilibrium. The most stable distribution is that with the highest possible entropy or disorder. According to *Boltzmann* (1905), there is a statistical relation between entropy, S, and probability, P.

$$S = k \ln P \tag{3.1}$$

where k is the *Boltzmann* constant; this is equal to R/N, where R is the gas constant, 1.987 cal/mole/deg, and N is *Avogadro*'s number. The molal entropy change, ΔS_1, of the first gas is given by:

$$\Delta S_1 = \frac{R}{N} \ln P_1$$

For the second gas:

$$\Delta S_2 = \frac{R}{N} \ln P_2$$

The condition for perfect mixing requires that P_1 be equal to P_2, and that ΔS_1 be equal to ΔS_2. Then for $\frac{1}{2}$ mole of oxygen to mix with $\frac{1}{2}$ mole of nitrogen to form a mixture with a total of N molecules:

$$\Delta S_1 = \Delta S_2 = 1.987 \times 2.303 \times 0.301 = 1.375 \text{ entropy units.}$$

The value of 1.375 entropy units is the entropy of mixing $\frac{1}{2}$ mole of each gas at constant temperature and pressure. The number 2.303 is the conversion factor for common to natural logarithms, and 0.301 is the logarithm of 2. In nonstatistical thermodynamics, the formula for the relationship between entropy and volume of an ideal gas is:

$$\Delta S = R \ln \frac{V_2}{V_1} \tag{3.2}$$

This represents the increase of entropy which occurs when a gas is expanded isothermally from an initial volume V_1 to a final volume V_2. The value is independent of path, process or mechanism, and depends only on the initial and final states. Thermodynamic functions depend only on changes of state rather than on paths or mechanisms. This is shown by the application of the Pfaffian differential equations of pure mathematics (*Caratheodory*, 1909; *Born*, 1948; *Joseph*, 1973). When a gas doubles its volume in a process of isothermal expansion, the value of the entropy change is:

$$\Delta S = 1.987 \times 2.303 \times 0.301 = 1.37 \text{ entropy units.}$$

When two liquids are perfectly miscible and form an ideal solution, the entropy of mixing $\frac{1}{2}$ mole of each of the liquids to form the mixture is:

$$\Delta S = R \ln 2 = 1.37 \text{ entropy units.}$$

Similarly, when 3 components of gas or liquid are mixed to form gas mixtures or perfect solutions, the value of ΔS is 2.18 entropy units when the mixture contains $\frac{1}{3}$ mole of each component. For a system of a components perfectly mixed in a liquid or vapor phase that contains $1/a$ moles of each of the components, the value of ΔS is $R \ln a$. Thus the entropy of any perfect mixture increases as the logarithm of the number of components. Therefore, perfectly miscible systems of many components have high entropies of mixing, and represent systems of high stability; the individual components tend to lose their individual properties shown in pure liquids. This is especially true of mixtures such as protoplasm, which contain many colloidal or macromolecular components in addition to water and inorganic ions (*Willstatter,* 1929; *Oparin,* 1938).

The Boltzmann Distribution

In the statistical distribution of black and white beans, there is no question of the probabilities being influenced by the colors or taste of the beans. The problem is treated as one of pure chance. In pure gases or in gas mixtures, however, the energy and entropy depend on the temperature and pressure, which determine the thermodynamic state. Energy is unequally distributed over the whole system of atoms and molecules. The problem of energy distribution is fundamental in kinetic theory and statistical mechanics. It is also fundamental in the theory of reaction rates. In general the problem is to express the frequency curve of the energies. In a system of n molecules, n_1 have the energy ϵ_1, n_2 have the energy ϵ_2, and so on for all the molecules in the system. The frequency curve expresses the numbers n_1, n_2, n_3 ... in relation to ϵ_1, ϵ_2, ϵ_3 ... In classical (or non-quantum) mechanics, the energy distribution is continuous. This gives rise to a continuous frequency curve, the classical *Boltzmann* distribution. In quantum mechanics there are quantized energy states with discrete energy levels. The problem requires the methods of quantum statistics, either the *Bose-Einstein* or the *Fermi-Dirac* statistics.

In the *Debye-Hückel* theory of interionic attractions, the method of classical statistics is sufficient to formulate an approximate equation of state for dilute electrolyte solutions. The electrostatic energy of the system depends on long-range attractions between the anions and cations. The interionic forces and energies are continuous rather than quantized functions. Many other problems, including the distribution of energies in gases, may be formulated according to the *Boltzmann* statistics.

Rigorous kinetic theory must account for two limiting kinds of phenomena. It must explain the statistical equilibrium described by entropy and the other thermodynamic functions. This requires the application of the calculus of probabilities to dynamic assemblies of molecules in which there is a statistical distribution of energies. It must also account for all the other mechanical properties of the system which are not thermodynamic. These include the frequencies of collisions, the diffusion constants, Brownian movement and 'noise', and thermal and electrical conduction. The theory of reaction rates includes consideration of collision rates as well as of certain thermodynamic properties. All properties are related to the thermal kinetic energies and to certain specific properties of the various substances.

The problem is similar to that of the equation of state for any condensed system. Any program for correlating the thermodynamic, kinetic and mechanical properties of physicochemical systems with the specific properties of the molecular components is beyond the scope of this chapter, the aim of which is to point out the relations between thermodynamic functions, reaction rates and statistical probabilities. The physiological aspect of the problem is to show how the thermodynamic properties may be correlated with the rates of metabolism or energy conversion.

In any assembly of atoms or molecules, each particle at any instant occupies an element of *phase space* described by generalized coordinates and momenta. The simplest particle, a monatomic molecule of mass m, occupies an element of phase space denoted by generalized coordinates and momenta. This element is denoted by dq_1, dq_2, dq_3, dp_1, dp_2, dp_3, where q denotes a coordinate and p denotes a component of the momentum. For example, the position of a helium atom is represented by three coordinates, x, y and z, and by three components of momentum, $m\bar{x}$, $m\bar{y}$ and $m\bar{z}$, where $\bar{x}$, $\bar{y}$ and $\bar{z}$ are the three velocity components. At any instant, the energy of the atom is determined by its coordinates and momenta, or in other words, by its position in phase space. Diatomic and polyatomic molecules require, in addition, angular coordinates and momenta.

In applying statistical laws, the total volume, V, of the phase space may be considered to be made up of a definite number, N, of very small volume elements, v_1, v_2, v_3 ... v_N. The ratios of these volume elements to V are denoted ω_1, ω_2, ω_3 ... ω_N. Therefore,

$$1 = \omega_1 + \omega_2 + \omega_3 \ldots + \omega_N \tag{3.3}$$

The number of molecules distributed over the N volume elements or *cells* is denoted as *n;* at every instant it is the sum of n_1, n_2, n_3 ... n_N, the numbers of molecules in each of the cells. The statistical distribution changes at every instant, but only under definite restrictions. One condition is that *n* is constant in a closed system:

$$n = n_1 + n_2 + n_3 \ldots + n_N \tag{3.4}$$

The energy of any atom at any instant is determined by its coordinates and momenta, which correspond to its position in phase space. The energies and positions are functions of time, but in a system at statistical equilibrium the total energy, E, is constant:

$$n_1 \epsilon_1 + n_2 \epsilon_2 + n_3 \epsilon_3 \dots + n_N \epsilon_N = E \qquad (3.5)$$

By designating the volume as V and the volume of a single cell as v_1, the number of atoms or molecules assigned to v_1 is n_1. The atoms in cell 1 are characterized by the energy ϵ_1. This energy is constant for the cell, but the number n_1 fluctuates. The fluctuations are subject to the restrictions of constant total number, n, and constant total energy, E. It is then assumed that the number of particles in each cell is proportional to the volume of the cell, but subject to the restrictions of constant total number, n, and constant total energy, E. (It is then assumed that the number of particles in each cell is proportional to the volume of the cell, but subject to the conditions of total number of particles and total energy.) The principles of probability are then applied to the distribution of particles in phase space. If the condition of constant E be temporarily omitted, the probability of a certain distribution, n_1, $n_2, n_3 \dots n_N$ is:

$$P_{(n_1, n_2, \dots n_N)} = \frac{n!}{n_1! \, n_2! \dots n_N!} \, \omega_1{}^{n_1} \, \omega_2{}^{n_2} \dots \omega_N{}^{n_N} \qquad (3.6)$$

where the ratio of the factorial numbers denotes the number of ways of arranging n particles in N cells of equal volume. A summation of all probabilities over all possible combinations $(n_1, n_2 \dots n_N)$ must be equal to 1. This agrees with the condition that

$$\Sigma \, \omega_1 = 1.$$

It is a property of the ratio

$$\frac{n!}{n_1! \, n_2! \dots n_N!}$$

that it shows a very sharp maximum for $n_1 = n_2 = n_3 \dots = n_N$.

Then a uniform distribution of particles in all the volume elements is by far the most probable distribution when the distribution of molecular energies is not taken into account. Uniform distribution of energy throughout the phase space implies uniform distribution of material particles.

In the general solution, three approximations are made, which are valid for very large values of n. First, the values for $n_1 \dots n_N$ are treated as continuous functions. Second, applying *Stirling*'s approximation to the equation of probability, P, the expression becomes:

$$\ln P = n \ln n - \Sigma n_1 \ln n_1 \tag{3.7}$$

The problem is now to determine the maximal value of P under the conditions of constant energy and constant number of particles. The method is as follows. The conditions for a maximal value of P are:

$$(1 + \ln n_1) \, \delta \, n_1 = 0$$
$$\Sigma \, \delta \, n_1 = 0$$
$$\Sigma \, \epsilon_1 \, \delta \, n_1 = 0$$

A general mathematical method exists for the solution of this type of problem. The method requires the use of *Lagrange*'s multipliers, α, β, which can be assigned any values. The second of the three equations is multiplied by α, and the third by β. Then

$$(1 + \ln n_1) + \alpha + \beta \, \epsilon_1 = 0$$

Let α and β be given values such that any two values of $\delta \, n_i$ and $\delta \, n_j$ become zero. Under the condition that P is to remain at its maximum, all possible values of $\delta \, n_1 \dots \delta \, n_N$ are admissible. The generalized solution includes zero values of these small variations. Hence the general solution requires that the coefficient of every term $\delta \, n_1 - \delta \, n_N$ must be equal to zero when α and β are chosen so that any two become zero. For the general distribution of the particles in phase space:

$$\ln n_1 + 1 + \alpha + \beta \epsilon_1 = 0$$
$$\cdots \cdots \cdots \cdots \cdots$$
$$\ln n_N + 1 + \alpha + \beta \epsilon_N = 0$$

Therefore:

$$n_1 = a \exp (-\beta \epsilon_1)$$
$$\cdots \cdots \cdots \cdots$$
$$n_N = a \exp (-\beta \epsilon_N)$$

where a is a constant. This system of equations represents the most probable distribution of energy. From the condition that $n_1 = n$, it is found that

$$a = \frac{n}{\Sigma \exp (-\beta \, \epsilon_1)} \tag{3.8}$$

and

$$n_1 = \frac{n \exp(-\beta \, \epsilon_1)}{\Sigma \exp(-\beta \, \epsilon_1)} \qquad (3.9)$$

By introducing the quantity f,

$$n_1 = \frac{n \exp(-\beta \, \epsilon_1)}{f} \qquad (3.10)$$

The quantity f is called the *partition function.* It depends on the nature of the frequency curve and it is related to the thermodynamic functions of the statistical assembly. The two preceding equations for n_1 give the classical *Boltzmann* distribution of kinetic energies in the system. An important step in the mathematical method is the proof that the coefficient β is independent of the nature of the molecules. The same multiplier enters the frequency distribution curve for all the particles in a mixture of two or more different molecular species. This entity is also independent of the kinds of energy $\epsilon_1, \epsilon_2 \dots \epsilon_N$. Small values of β indicate high probabilities that the molecules will have any given quantity of energy; this means a frequency distribution over a wide range of kinetic and potential energies. These properties are satisfied by the definition

$$\beta = \frac{1}{kT}$$

where k is the *Boltzmann* constant and T is the absolute temperature. Then the energy distribution law is formulated as:

$$n_1 = \frac{n \exp\left(\dfrac{-\epsilon_1}{kT}\right)}{f} \qquad (3.11)$$

Work Function and Probability

The statistical distribution of energy permits formulations of the thermodynamic functions. The expression for the number of molecules, n_1, having the energy ϵ_1, leads to the following result:

$$\ln n_1 = \ln n - \frac{\epsilon_1}{kT} - \ln f \qquad (3.12)$$

Then

$$n_1 \ln n_1 = n_1 \ln n - \frac{n_1 \, \epsilon_1}{kT} - n \ln f \qquad (3.13)$$

and

$$\Sigma \, n_1 \ln n_1 = n \ln n - \frac{E}{kT} - n \ln f \tag{3.14}$$

In this expression, E includes only that part of the total energy that is dependent on the temperature. It does not include the intrinsic bond energy, E_0 which represents the energy at absolute zero. The expression for ln P is:

$$\ln P - n \ln n - \Sigma \, n_1 \ln n_1 = \frac{E}{kT} \tag{3.15}$$

According to equation (3.1),

$$S + \frac{E}{T} + R \ln f \tag{3.16}$$

Thus in any process in a heterogeneous gas mixture or solution in which

$$\omega_1 = \omega_2 = \omega_3 \ldots = \omega_N$$

and in which

$$\epsilon_1 = \epsilon_2 = \epsilon_3 \ldots = \epsilon_N$$

$$f = \Sigma \, \exp \left(\frac{-n_1 \, \epsilon_1}{kT} \right)$$

and

$$dS = k \, d \ln P = 0 \tag{3.17}$$

This shows that any perfect solution is a homogeneous mixture, the entropy of which is maximal at equilibrium. In a heterogeneous system:

$$\omega^{1'} = \omega_2{}' = \omega_3{}' \ldots = \omega_N{}'$$
$$\omega_1{}'' = \omega_2{}'' = \omega_3{}'' \ldots = \omega_N{}''$$

where $\omega_1{}'$ denotes a volume element of phase 1, and $\omega_1{}''$ denotes the corresponding element of phase 2. Then:

$$\Delta S_c = R \, \Sigma \, \exp \frac{n_1 \, (\epsilon_1{}'' - \epsilon_1{}')}{kT}$$

where the total number of molecules $\Sigma\, n_1$ is treated as a constant, and where $\bar{S_c}$ denotes configurational entropy. Then the entropy change implies an increase of entropy in each volume element of phase 2. A positive value implies an increase of entropy. Mixing of two phases would then imply a decrease of entropy. The coexistence of two phases at constant energy and temperature therefore depends on the values of $(\epsilon_1{}'' - \epsilon_1{}')$ for each volume element, as applied to substance 1; the same considerations apply to every substance. As $\Delta\epsilon_1$, the energy difference, approaches zero for each element ω, the ratios of P'' to P' and of f'' to f' approach 1.0. Therefore the two mixtures can no longer coexist as different phases. The expression

$$\Delta S_c = R \ln \frac{P''}{P'}$$

thus illustrates an application of *Carnot*'s principle, according to which there are inaccessible states in the neighborhood of any given accessible state (*Caratheodory*, 1909; *Born*, 1948). Thus a homogeneous solution (phase 1) is inaccessible in the neighborhood of a second homogeneous solution (phase 2) of the same components when the two solutions have different configurational entropies. Conversely, the second phase is inaccesible when its formation involves a change of configurational entropy at constant temperature and energy.

From equation (3.16) it follows that

$$E - TS = -RT \ln f + E_0 \tag{3.18}$$

and that A, the '*Helmholtz* free energy or work function' is given by:

$$A = -RT \ln f + A_0 \tag{3.19}$$

From this relationship, the energy and entropy are formulated by differentiation:

$$\frac{\partial A}{\partial T} = -S$$

$$S = R \ln f + RT \left(\frac{\partial \ln f}{\partial T}\right) \tag{3.20}$$

Energy and entropy are related by the formula, $E = A - TS$, and therefore:

$$E = RT^2 \left(\frac{\partial \ln f}{\partial T}\right) + E_0 \tag{3.21}$$

Homogeneous Equilibrium

Homogeneous chemical reactions are characterized by their equilibrium constants and standard free energy changes. For example, in the vapor phase, the reaction

$$H_2 + \frac{1}{2} O_2 = H_2O$$

shows a very high yield of water vapor in equilibrium with very low pressures of hydrogen and oxygen gases. The equilibrium constant, K, is related to the pressures by the law of mass action:

$$\frac{p_{H_2O}}{p_{H_2} \times p_{O_2}^{1/2}} = K$$

and K is related to the standard free energy change by the formula:

$$\Delta G^0 = -RT \ln K$$

where ΔG^0 is the change of standard *Gibbs* free energy.

In the above reaction there is a partition function for each of the reactants and for the product, H_2O. These functions change during the course of the reaction. The change is expressed by the summation $\Sigma \, a_1 \ln f_1$, where *a* refers to the number of moles of each substance. It is positive for the product and negative for the reactants. For H_2O it is 1, for hydrogen gas it is -1, and for oxygen it is $-\frac{1}{2}$.

The total free energy change depends on two effects: the change of partition functions and the change of enthalpy. The partition functions are related to the distribution of energies among the molecules. The change of free energy, ΔG, may be regarded as the sum of the two effects. The part of the free energy which is temperature dependent and related only to the distribution of thermal energy is given by:

$$\Delta G_k = \Delta H_k - T\Delta S$$

where ΔH_k is the kinetic and mechanical part of the enthalpy change. Since

$$\Delta H_k = \Delta E_k + P \, \Delta V$$

$$\Delta H_k = RT^2 \, \Sigma \, n_1 \left(\frac{\partial \ln f_1}{\partial T} \right) + P \, \Delta V.$$

The reaction is attended by changes of chemical bond energies, expressed by ΔH_b. The part of the total change of free energy related to this is expressed by

$$\Delta G_b = \Delta H_b.$$

The total change of free energy is

$$\Delta G = \Delta G_b + \Delta G_k.$$

Therefore

$$\Delta G = \Delta H_b - RT \, \Sigma \, (a_1 \ln f_1).$$

Now the partition functions refer to all forms of thermal energy in the system, including translational energy. The concentrations are inversely proportional to the volume. The volume elements ω_1, ω_2 ... ω_N, are proportional to the total volume. The free energy equation may be modified to express the relation of ΔG to the concentrations. Then

$$\Delta G - RT \, \Sigma \, (a_1 \ln f_1) + RT \, \Sigma \, (a_1 \ln c_1) = 0$$

where

$$\ln K = \Sigma \, (a_1 \ln f_1) - \frac{\Delta H}{RT}$$

Applying this relation to the given equilibrium,

$$K = \frac{f_{H_2O}}{f_{H_2} \times f_{O_2}^{1/2}} \exp \left(\frac{-\Delta H}{RT} \right)$$

The equilibrium thus depends on the partition functions and on the standard change of enthalpy. The ratio of the partition functions depends on the ratio of the number of reacting molecules to the number of molecules that are formed. In the formation of water, two hydrogen molecules react with one oxygen to form two molecules of water. The reaction represents a condensation of three molecules to form two. In general, when a large number of molecules react to form a smaller number, the ratio of the partition functions tends to be less than 1. This operates to decrease the yield. However, a high negative value of ΔH, which represents a high decrease of enthalpy, can produce a highly negative value of ΔG^0, a high value of K and a high yield of the product. This factor can overcome a decrease of entropy which might occur when few molecules are formed from many. Endothermic reactions, on the other hand, are those which require an increase of entropy to compensate for an increase of enthalpy. In such reactions, equilibrium constants greater than 1.0 result from the values of the partition functions. These implications of statistical thermodynamics are expressed in the following formulations:

$$K = \exp\left(\frac{\Delta S}{R}\right) \exp\left(\frac{-\Delta H}{RT}\right)$$

Then the equilibrium changes with the temperature according to the formula:

$$\frac{\partial \ln K}{\partial T} = \frac{\Delta H}{RT^2}$$

Chemical equilibrium in homogeneous systems may involve large changes of bond energy, expressed as ΔH. The thermal changes involved in solubility, miscibility and other kinds of phase equilibrium are usually of a smaller magnitude. Nevertheless these different phenomena may be quantitatively reduced to similar formulations expressed in relation to the changes of entropy and enthalpy; they are then related to the partition functions, which are factors in the accessibility or inaccessibility of various states of aggregation in various kinds of physicochemical systems.

In the formation of water vapor from hydrogen and oxygen gases, $\Delta S^0{}_{298.1}$ is -11.1 entropy units; $\Delta H^0{}_{298.1}$ is $-57,820$ cal. The corresponding value of $\Delta G^0{}_{298.1}$ is $-54,507$ cal. The equilibrium constant calculated from ΔG^0 is of the order of 10^{40} at 25 °C. In this case K is estimated from the calorimetric values of ΔH and ΔS. When it is possible to evaluate K directly, ΔS and ΔH may be evaluated from equilibrium data.

Reaction Rates and Temperature

A reversible equilibrium depends on two simultaneous reactions progressing in opposite directions. The simplest case is that of a monomolecular reaction of the type, $A \rightleftharpoons B$. At equilibrium:

$$\frac{c_B}{c_A} = K$$

where the equilibrium constant, K, may be expressed as the ratio of two rate constants, k_1 and k_2. The rate of formation of B is given by $k_A c_A$. The rate of the reverse reaction is expressed as $k_B c_B$. At equilibrium the two rates are equal and opposite. Then, the *net* rate of a reversible reaction is zero. This is characteristic of all reaction rates in a system of any number of reactants and products in which all the reactions are reversible. At equilibrium such a system may be treated as one component. The simplest example would be a mixture of two substances, A and B, either in a steady state or at equilibrium. In either case the state is at statistical equilibrium. It follows that K is equal to k_2/k_1. Therefore:

$$\frac{d \ln K}{dT} = \frac{d \ln k_2}{dT} - \frac{d \ln k_1}{dT}$$

It follows that:

$$\frac{d \ln k_2}{d (1/T)} - \frac{d \ln k_1}{d (1/T)} = \frac{-\Delta H}{R}$$

By integration,

$$\ln K = C - \frac{\Delta H}{RT}$$

where C is the integration constant. Hence:

$$\ln \frac{k_2}{k_1} = C - \frac{\Delta H}{RT}$$

On the basis of this, *Arrhenius* proposed an equation for the rate constant, k, of any reaction:

$$\ln k = C - \frac{E_a}{RT}$$

where E_a is the 'activation energy'.

Modern kinetic theory relates E_a to a quantity of energy required to activate the molecules. This may be considered to be the minimal vibrational energy, or the minimal electronic excitation energy required for chemical change. The number of molecules in this state depends on the temperature, and is proportional to

$$\exp\left(-\frac{E_a}{RT}\right).$$

The number of activated molecules depends on the ratio of the activation energy to the absolute temperature. A low ratio favors high reaction rates, whereas a high ratio corresponds to a slow reaction. Many reaction rates of biological importance are approximately doubled with each $10°$ increase of temperature. An increase of this magnitude corresponds to an activation energy of about 12,000–13,000 cal.

It is clear from thermodynamics and from statistical mechanics that distribution constants and reaction rates are functions of temperature. Linear relations are obtained when ln K and ln k are plotted against the reciprocal of the absolute temperature. It is essential to bear this in mind when biological systems are considered. It is impossible to change the temperature of cells and tissues without simultaneous changes of all equilibrium constants and all reaction rates. These include the rates of oxygen consumption and energy conversion, which are functions of temperature.

When metabolizing cells are experimentally cooled, a number of simultaneous changes may be observed to occur, including redistributions of water and electrolytes. The cause of these changes is the cooling of the cells; heat is transferred to the environment. Like any other aggregation of molecules, the cells lose thermal kinetic energy and their entropy is lowered. The remaining energy is redidistributed according to statistical principles, causing changes in all thermodynamic functions and reaction rates. All the parameters which characterize the system are functions of one independent variable, the temperature. Temperature is here the primary function, and metabolic rates and equilibrium constants are secondary. Therefore, it cannot be said that any metabolic rate or mechanism is the independent variable or the cause of the observable phenomena.

Cells and tissues are heterogeneous systems with a fixed number of components and phases. In perfectly steady states (homeostasis), the physicochemical system is invariant. The distribution of the components depends on the characteristic thermodynamic functions such as standard free energies and entropies, and the steady reaction rates depend on the frequency curves of energy distribution among the bulk and microcomponents. In systems with a high degree of internal organization, the number of equations of condition approaches the number of variables. Only by this supposition can one explain the integrated behavior which is characteristic of living organisms. If this supposition is not made, then each equation of condition or constraint must be replaced by an arbitrary number of hypothetical ad hoc mechanical entities. A large number of such entities would be required to maintain order in the system and to account for coherent behavior. The thermodynamic and statistical equations of condition and reaction rates are statements of maximal probabilities, and apply to heterogeneous systems in general, whatever may be the states of order or disorder.

Chapter 4

Brownian Movement and Diffusion

In heterogeneous systems there are processes of diffusion and exchange in all physiological states in which the free energy is not minimal. Redistributions of the diffusible components occur in the direction in which the chemical potentials decrease. For electrolyte components 'chemical potential' refers to the sum of the potentials of anions and cations; this depends on the type of electrolyte (binary or ternary). The distributions become stable under steady-state conditions of minimal free energy and constant chemical potentials. The latter condition applies in general only to uniformly homogeneous systems with no surfaces of discontinuity and to systems in which there are no exothermic metabolic reactions. It is therefore not applicable without qualifications to biological systems. The qualifications apply to the necessary conditions for invariance in actively metabolizing systems of cells and tissues (*Joseph*, 1971a, b; 1973).

The equilibrium distribution of chemical substances is discontinuous in systems that contain more than one homogeneous phase; such systems do not show the property of perfect mixing which is found in all homogeneous solutions. Under these conditions, according to statistical principles, the distribution of energy and entropy is also discontinuous and depends on the partition functions, which are factors in all equilibrium and rate constants (chapter 3). In biological systems, the coexistence of discontinuous phases in various accessible and inaccessible states implies high configurational free energy and low configurational entropy in cells and tissues, in which water is in a structured solid or semi-solid state of aggregation (*Schrodinger*, 1944; *Joseph*, 1973).

The diffusion of material substances within homogeneous liquid phases is governed by *Fick*'s law, according to which the rate of flow of material is proportional to the gradient, or rate of change of concentration per unit of distance. The proportionality constant depends on a standard rate of diffusion characteristic of each solvent and solute. For inorganic ions the *diffusion constant*, D, is proportional to the electrical mobility or ionic conductance. The net rate of flow is then proportional to the surface area, the diffusion constant and the concentration gradient. If at any point x, there is a concentration, c, and at a neighboring point, x + dx, the concentration is c + dc, then the homogeneous concentration gradient is expressed by the coefficient

$$\frac{dc}{dx}.$$

Diffusion in a homogeneous system occurs as long as the concentration gradient exists. The rate is measured by dn, the number of molecules or ions which diffuse across a cross-sectional area, A, in time dt. Then

$$dn = -DA\left(\frac{dc}{dx}\right)dt$$

where D is the diffusion constant. A general solution of the problem of diffusion requires establishing the integral relation:

$$c = f(x, t).$$

The linear diffusion is measured by determining the concentration c at any point x at time t. At equilibrium c is independent of x and t; it is then invariant throughout the homogeneous phase. The general relation between the rate at which the concentration c changes with time at any point x is expressed by the equation:

$$\left(\frac{\partial c}{\partial t}\right)_x = D\left(\frac{\partial^2 c}{\partial x^2}\right)_t \tag{4.1}$$

This is a second order differential equation, and the solution in any case depends on the specific conditions of the problem as expressed by the integration constants. The equation states that the rate at which the concentration changes with time at any point x is proportional to the rate of change of the concentration gradient along the x axis at a given time, t. A general solution of the problem depends on the initial conditions, in which c is a known function of x at the initial time. The condition of invariance of c at all points is that the right-hand term of the differential equation be zero. This requires either that D or the second differential coefficient of c with respect to x be zero. In the latter case, the concentration gradient would be uniform and independent of x and t.

Density in Phase

The mathematical formulations of classical Newtonian non-quantum mechanics culminated in the works of *Laplace, Lagrange, Gauss, Hamilton* and many other mathematicians of the 18th and 19th centuries. This work, when combined with the theories of probability and games of chance initiated by *Pascal* and *Fermat* in the 17th century, made possible the subsequent theory of statistical mechanics of *Gibbs* (1901), and other developments due to *Boltzmann* and *Einstein*. The mathematical theory of Brownian movement, diffusion and osmotic pressure as developed by *Einstein* (1905–1908) was likewise based on 19th century formulations in theories of heat, thermodynamics and statistical mechanics. These formulations in many ways were similar to the foundations of *Gibbs'* methods. A collection of five of *Einstein*'s papers edited by *R. Furth* has been republished (1926). Subsequent developments of 20th century statistical mechanics required the incorporation of quantum mechanics, as developed in the *Fermi-Dirac* and *Einstein-Bose* statistics. These non-Newtonian developments, although they have revolutionized theoretical physics, have required no essential revisions in the 19th century formulations of *Hamilton, Boltzmann, Gibbs,* or in the classical theories of Brownian movement and diffusion.

In a dynamic system of particles, the total energy of which is denoted by (T + V), the Hamiltonian function (H, p) is expressed as a function of the generalized coordinates, q ... and the generalized momenta, p ... of all the particles.

$$T + V = H(p, q).$$

T, the total kinetic energy, is a function both of the coordinates (q) and the generalized velocities ($\dot{q}_1, \dot{q}_2 ... \dot{q}_n$). The generalized momenta are defined by equations of the form

$$p = \frac{\partial T}{\partial \dot{q}}$$

In the canonical form, the equations of motion are written as:

$$\dot{q} = \frac{\partial H}{\partial p}; p = -\frac{\partial H}{\partial q}$$

Conservation of energy implies that changes of (T + V) and H are equal to zero. Applying the equations of motion in the canonical form, the distribution of particles in *phase space* is described by a function $f(t, q_1, q_2 ... p_1, p_2 ... p_n)$ of all coordinates and momenta and of time.

The element f dp dq is the probability (*Gibbs'* density in phase) that the system will occur at time t in a given element $dp\ dq = dp_1 ... dp_n\ dq_1 ... dq_n$. Density in phase may be considered to represent the density of a fluid in a pq 'phase space' of 2n dimensions, where 2n is the number of generalized coordinates and momenta. The relation of f to t is expressed by:

$$\frac{df}{dt} - [H, f] = 0 \tag{4.2}$$

where [H, f] denotes the '*Poisson* bracket', expressed by

$$[H, f] = \Sigma \left(\frac{\partial H}{\partial q_k} \frac{\partial f}{\partial p_k} - \frac{\partial H}{\partial p_k} \frac{\partial f}{\partial q_k} \right).$$

Liouville's theorem expresses the equivalence of equation (5.2) with the result

$$\frac{df}{dt} = 0.$$

Thus f, the density in phase, is constant along any path of the assembly in phase space. This expression for the integral of the canonical equations is denoted by *Gibbs* as 'conservation of density in phase'. Thus the integral ∫ f dp dq is independent of time. It is therefore a conservative function.

The canonical distribution of generalized coordinates and momenta in *Gibbs'* treatment corresponds also to the *Maxwell-Boltzmann* distribution when

$$f = \exp (\alpha - \beta E)$$

and

$$H(p, q) = E.$$

For systems in thermal equilibrium, $\beta = 1/T$. The phase integral then becomes:

$$\int \exp (\alpha - \beta E) \, dp \, dq$$

The methods applied to the trajectory of a system of particles in 2n dimensional phase space (p, q) were developed independently by *Einstein* (1905–1908) to describe Brownian movement. The formulation permits the direct observation of the coordinates and motions of microscopically observable particles, confirming many of the theoretical calculations of kinetic theory relating to mass and energy of the particles. Study of the fluctuations of positions of the colloidal particle in the field of the microscope led, for example, to the direct calculation of *Avogadro*'s number. The theory also permits calculation of diffusion constants from other molecular parameters such as molecular weights and viscosities.

In biological systems that contain protein molecules and colloidal aggregates of submicroscopic dimensions (100–1,000 Å) Brownian movement causes fluctuations of energy, entropy and heat capacities in intracellular submicroscopic volumes of the coordinate (q) space. These fluctuations of the thermodynamic functions would depend on submicroscopic division of the space within discrete physicochemical phases determined by the chemical morphology of cellular cytoplasm and nucleus. Ultimately the fluctuations depend on the chemical composition of the macromolecular polyelectrolytes and colloidal aggregates of the intracellular phases. These are regions of low entropy and high order in which the stabilizing intermolecular attractions are sufficiently strong to overcome the molecular chaos that is the cause of disorder, Brownian movement, diffusion, high entropy and partition coefficients. The latter functions depend on the distribution of entropy among all the atoms and molecules of the assembly. 'Orderly biological processes are unthinkable without presupposing structure, and it is therefore out of the question that any living constituent of protoplasm could consist of structureless, fluid, independently displaceable particles' (*Frey-Wyssling*, 1953).

If it is necessary to presuppose structure to account for orderly biological processes of growth, metabolism and reproduction, it is also necessary to consider orderly processes of diffusion, reversible and irreversible transport, heat conduction and respiratory metabolism. These depend on the molecular kinetic parameters that are responsible for Brownian movement, diffusion and fluctuations of the thermodynamic functions. Statistical molecular motion is the basis of irreversible biological processes, and may be related to the 'standard errors' which are universally found. Molecular disorder and high entropy within the less highly organized regions of the coordinate space are attributable to the fluctuations due to chance. Physicochemical changes of state and mixing phenomena are related processes which may be related to statistical laws.

Brownian Movement

Einstein's first paper on the subject of Brownian movement, osmotic pressure and diffusion appeared in 1905 under the title *On the Movement of Small Particles Suspended in a Stationary Liquid Demanded by the Molecular-Kinetic Theory of Heat.* It is supposed that a number of suspended particles in a state of dynamic equilibrium are irregularly dispersed in a liquid. A force K is assumed to act on the single particles, being exerted always in the direction of the x axis. If v is the number of particles per unit volume, the condition for equilibrium is found to be:

$$-K + \frac{RT}{N} \frac{\partial v}{\partial x} = 0 \qquad\qquad (4.3)$$

or

$$K - \frac{\partial p}{\partial x} = 0 \qquad\qquad (4.4)$$

where p is the osmotic pressure deduced from the molecular kinetic theory of heat (*Einstein,* 1926). Thus the force K is related to osmotic forces as derived from kinetic theory.

Equations 6.2 and 6.3 are used to determine the diffusion coefficient of the suspended substance in a liquid of viscosity η. If the particles are spheres of radius r, then the velocity of a particle in the direction of the x axis is given by *Stokes'* law:

$$V_x = \frac{K}{6 \pi r \eta}$$

Per unit area, the number of particles diffusing in unit time in the direction of decreasing osmotic presure is then

$$\frac{K}{6 \pi r \eta}$$

Denoting the diffusion coefficient as D, the number of particles per unit area is

$$-D \frac{\partial v}{\partial x}$$

Under the condition of dynamic equilibrium,

$$\frac{K}{6 \pi r \eta} - D \frac{\partial v}{\partial x} = 0.$$

Thus the diffusion constant depends only on the radius of the spherical particles and on the viscosity of the liquid.

To account for the irregular motions of the particles in a state of dynamic equilibrium (Brownian motion), a more detailed consideration of the motion of individual particles is required. A time interval of definite magnitude τ is introduced; this is very small compared to the total time t that is experimentally observed, but is of sufficient magnitude that the movements of a particle in two consecutive intervals τ can be considered to be independent of each other.

In the time interval τ the x coordinates of the particles increase by Δ, which has a different value (positive or negative) for each particle. A certain probability law holds for Δ whereby the number of particles between Δ and $(\Delta + d\Delta)$ is expressed by:

$$dn = n \, \varphi (\Delta) \, d\Delta$$

where $\varphi (\Delta)$ fulfills the condition for maximal probability.

Solution of the equation yields the differential equation for diffusion:

$$\frac{\partial c}{\partial t} = D \frac{\partial^2 c}{\partial x^2}$$

When the mean displacement along the x axis in the total time interval t is expressed as $\bar{\Delta}$, it is found that

$$\sqrt{\bar{\Delta}^2} = \sqrt{2\,D\,t}$$

When D is eliminated

$$\bar{\Delta}^2 = \sqrt{t}\ \sqrt{\frac{RT}{3\,\pi\,r\,\eta}}$$

The corresponding expression for *Avogadro*'s number N becomes

$$N = \frac{1}{\bar{\Delta}^2}\frac{RT}{3\,\pi\,r\,\eta}$$

The quantity $\bar{\Delta}^2$ represents the root mean square of the mean displacements during equal time intervals. The following results are obtained for the arbitrary values at 17 °C, when N is taken as 6×10^{23} and η is assigned the value 0.0135.

$$\bar{\Delta}^2 = 8 \times 10^{-5} \text{ cm} = 0.8\ \mu\text{m}$$

when the diameter of the particles is taken as 0.0001 mm. This yields a value of about 6 μm for a time interval of 1 min.

Fluctuations

The root mean square of the displacement $\bar{\Delta}^2$ from which estimations of *Avogadro*'s number may be made, is also related to other observable phenomena connected with the Brownian movement. For example, the fluctuations of the number of suspended particles within a given small volume studied over a period of time may also be used in the calculation of N (*Smoluchowski,* 1912). Phenomena which, according to *Boltzmann*'s relation between probability and entropy, would be irreversible, may actually be reversible for very small volumes of a suspension in communication with a large volume. Thus the process of diffusion, ordinarily irreversible on the macroscale may be observed to be reversible when only certain very small volumes are considered. This requires a general reconsideration of *Carnot*'s principle (second law of thermodynamics), which may not always be applicable to small numbers of particles in microscopic samples. The fluctuations of these numbers are determined by the theory of Brownian movement.

In studying the fluctuations, there are two necessary kinds of observations: first the average number of particles in a given small volume, and second, the probable value over a time interval. *Smoluchowski* derived the following formula for the probability:

$$P(n) = \frac{\exp{(-\nu)}\, p^n}{n!}$$

where ν is the average number of particles in the small volume, and p is the osmotic pressure, RTc. The average fluctuation is defined as $\bar{\delta}$, where

$$\bar{\delta} = \frac{n - \nu}{\nu}$$

This is always taken as positive. Two limiting values of $\bar{\delta}$ can be evaluated: (a) for small values of ν

$$\bar{\delta} = \frac{2}{\sqrt{\pi \nu}}$$

(b) for large values of ν

$$\bar{\delta}^2 = \frac{1}{\nu}$$

Svedberg (1928) and *Westgren* (1916) tested these and similar formulae with the following results:

Particles per μm^3	ν	δ		
		observed	calculated	
10	1.545	0.660	0.656	Svedberg
72.5	2.168	0.539	0.527	Westgren

Compressibility was also studied and compared with that of a perfect gas. As a first approximation, osmotic pressure conforms to the ideal *van't Hoff* law, a result also obtained by *Einstein* (1926). The fluctuations of osmotic pressure indicate statistical deviations from thermodynamic equilibrium; this is also found for other thermodynamic functions which fluctuate for small values of ν in very small volumes.

Smoluchowski's theory of fluctuations also yields other interesting results. When the time of observation is extended over rather long time intervals, two related phenomena may be studied. First, the average length of time during which a certain number of particles, n, remains constant, and second, the time interval necessary for a certain number to return to an initial value. This is the time necessary for the recurrence of a given state. The first time is designated as the average duration of n. It is given by:

$$T_n = \frac{t}{1 - P(n, n)}$$

where t is the duration for a given period.

The average time of return of n is

$$\theta_n = T \frac{1 - P(n)}{P(n)}$$

The following values of T_n and θ_n have been calculated by Smoluchowksi (Svedberg, 1928):

Fluctuations in a gold sol

n	T_n, sec observed	calculated	θ_n observed	calculated
0	2.37	2.26	9.35	8.52
1	2.31	2.38	4.81	4.86
2	2.11	2.12	6.32	6.23
3	1.92	1.89	12.1	12.4
4	1.80	1.72	28.6	32.1

ν = 1.55 (average); t = 1/39 min.

The results for the time of return θ_n are of the order of 30 sec or less for small values of n. However, θ_n increases rapidly for higher values of n. The value of the time of return is found to be 28 min when n is 7, when the average value of n is 1.55. When the value of n is 17, and the average value is 1.55, the calculated value of the time of return is about 50,000 years. Even in the case where n = 17, the diffusion of the particles is not absolutely irreversible. The time of return is, however, so great as to be unobservable. It is evident that thermodynamic reversibility and irreversibility are not *absolute* concepts. From the theory of probability, statistical processes can only *approach* states of absolute reversibility or irreversibility.

Optical Effects

Fluctuations of the number of elementary particles in a small fraction ω V of the whole volume of a gas or liquid give rise to fluctuations of density and of osmotic pressure. Along with these, there occur not only numerous kinds of optical phenomena — light scattering, transmission polarization and refractive indices — but also simultaneous fluctuations of thermodynamic functions: entropy, heat capacities and chemical or electrical potentials. On a small scale, irreversible properties are observed which almost never occur in large masses or volumes of homogeneous physicochemical systems or solutions. Similarly, sub-microscopic fluctuations of thermodynamic properties in heterogeneous bio-logical systems, although usually not directly observable, may be conceived to give rise to experimentally observable or operational properties which may be, at present, poorly understood but nervertheless of conceivable importance. These may give rise to observable optical phenomena in the realms of light scattering, polarization and light microscopy. These would depend on frequencies in the visible, ultraviolet or infrared regions of the spectrum.

When a beam of light traverses a vacuum or a perfectly structureless medium of constant refractive index, there is no scattering, refraction or diffraction in any part of the medium. In a vacuum or at very low gas pressures, the transmission, T, approaches 100%, and the optical density $(100 - T)$ approaches zero. When, however, the medium is heterogeneous and contains dust particles or certain kinds of vaporized liquids, there may be scattering of the light in all directions, a decrease of transmission that depends on the wavelength, and a corresponding increase of optical density. These effects are proportional to the length of the path through the inhomogeneous medium.

Lord *Rayleigh* (1871) developed a formula for the intensity of scattered light which shows

$$\text{intensity at angle } \theta = A^2 \frac{(D' - D)^2}{D^2} (1 + \cos^2 \theta) \frac{n \, \omega^2 \, V^2}{\lambda^4 \, r^2}$$

where A^2 is the intensity of the incident light, D' and D the optical densities of the particles and the medium in which they are suspended, n the number of particles, λ the wavelength of the light and r the radius of the particles. Thus the intensity of the scattered light at any angle θ, referred to the direction of propagation, is inversely proportional to the fourth power of the wavelength.

Light in the red or orange regions of the visible spectrum tends to be transmitted and blue or violet light tends to be scattered. In this way Lord *Rayleigh* accounted for the blue tint of the sky. The theory also accounts for the redness of the sunset; the red part of the spectrum is transmitted, as the short wavelengths are scattered.

The same theory is able to account for many other phenomena of light scattering, for example the blueness of tobacco smoke from a cigar or pipe, or the blueness of smoke from a log fire. Light scattering is due to a high density of colloidal particles suspended in the surrounding air. The stability of any such suspension depends on Brownian motion of the particles. Vapors of various organic liquids may also show light scattering of the blue and violet rays. On the other hand, light of long wavelengths in the infrared are transmitted through fogs and water vapor under conditions in which visible light is absorbed. When a properly sensitized plate is exposed to transmitted light in the infrared, photographic development is found to show details and effects that are invisible under conditions in which the visible light is absorbed or scattered.

Light scattering in the atmosphere does not necessarily depend only on the presence of suspended dust particles or other kinds of inhomogeneous substances. It can also be explained by statistical inhomogeneities due to spontaneous fluctuations of density and pressure in air, which is a mixture mostly of oxygen and nitrogen. Thus in a given volume of gas a small volume element ω V represents a small part of a given large volume V. If the volume is divided into two parts proportional, respectively, to ω and $(1 - \omega)$, then

$$\sqrt{\bar{n}_1{}^2} = \bar{n}_1$$

where $\bar{n}$ is the mean square deviation. This corresponds to the density fluctuation in a small volume, based on the *Boltzmann* statistics. Other fluctuations are related to density fluctuations in energy, pressure, heat capacity, and so forth.

Molecular Weights from Light Scattering

The methods developed by *Smoluchowski* (1912) and *Einstein* (1926) for statistical fluctuations in non-ideal colloidal solutions may be applied to the theory of light scattering and related to the molecular weights of large polymers. The methods depend on consideration of the fluctuations of the refractive index, as related to fluctuations of density and osmotic pressure. These occur within volume elements which are small compared to the wavelength of the incident light. It is unnecessary to consider the fluctuations that occur in the pure solvent. The light scattering due to the dispersed polymers gives rise to changes in the turbidity; these are related to fluctuations of concentration. When the concentrations in the small volume elements fluctuate, due to Brownian movement, there are observable changes in the polarizability α.

$$\Delta\alpha = \frac{\Delta\epsilon \; \omega \; V}{4\,\pi}$$

where $\Delta\epsilon$ is the displacement of optical dielectric constant for ω V compared with the average for V. It may then be shown that

$$\sqrt{\overline{\Delta\epsilon^2}} = \left(\frac{\partial\epsilon}{\partial c}\right)^2 \overline{\Delta c^2}$$

and

$$R_\theta{}^0 (1 + \cos^2 \theta) = K^* c\, RT \left(\frac{dp}{dc}\right)$$

where p is the osmotic pressure, θ is the angle of the scattered light, R is the gas constant and T is the absolute temperature. The quantity $R_\theta{}^0$ is known as the *Rayleigh* ratio. It is related to K^* and the molecular weight of the polymer by the following relationships:

$$R_\theta{}^0 = K^* (1 + \cos \theta)\, Mc$$

$$\frac{2\,\pi^2\, n_\theta{}^2 \left(\dfrac{d\bar{n}}{dc}\right)^2}{N \lambda^4} = K^*$$

N denotes *Avogadro*'s number, and $\bar{n}$ the refractive index. The change of $\bar{n}$ with c is expressed by the differential coefficient. Accordingly K^* and $R_\theta{}^0$ can be evaluated from experimental data and the molecular weights from the foregoing equations (*Debye*, 1947; *Flory*, 1953).

By means of another quantity, H, the molecular weight may be related to the observed turbidity τ of a colloidal suspension. Under certain limiting conditions

$$\tau = H\, c\, M$$

where H depends in a rather complicated way on wavelength, refractive indices and the change of refractive index with concentration.

As an example of the application of optical methods to the determination of the molecular weight of an important protein, some of the physicochemical properties of tobacco mosaic virus were studied (*Boedtker and Simmons*, 1958). The molecular weight was found to be 39.5×10^6. The molecular kinetic units could be described as rods of 2,900–3,200 Å in length. Values of the molecular weights of other viruses have been estimated as 15,000,000–45,000,000 (*Cohn and Edsall*, 1943). Other types of proteins studied by the methods of sedimentation equilibrium and sedimentation velocity are of the order of 17,600 to 6.76×10^6 (*Svedberg*, 1938). These values refer to various kinds of oxygen carriers of blood (hemoglobins, hemocyanins and chlorocruorins).

Spreading

When solutions or chemical substances in other states are injected into blood or other biological fluids or tissues, the normal standard states of equilibria are disturbed. When for example, certain dyes are injected into blood, there may be an initial period in which diffusion may lead to a rather rapid equalization of concentration throughout the entire volume of blood. In such a rapid process, only a homogeneous aqueous phase is involved. The requirement for this is that the equilibration period must be very rapid in comparison with the time required for diffusion into connective tissue, basement membrane, or intracellular phases. In the later phases of the process, dyes or other substances migrate into many different phases of a varied heterogeneous system. In this case it seems preferable to denote the process as 'spreading' rather than as 'diffusion'.

In earlier sections dealing with diffusion through homogeneous liquid media or solutions, it has been possible to relate the process to Brownian movement as determined by viscosity, particle size and 'concentration gradients'. This may not be possible in heterogeneous biological systems, which are characterized by many discontinuous boundaries. For example, in the ground substance of connective tissues, the submicroscopic vacuoles which contain the bulk of the tissue fluid as a water-rich phase, the total surface of the walls of the vacuoles is of a very high order of magnitude (*Gersh and Catchpole*, 1960; *Engel et al.*, 1961). At these walls the surfaces are lined with various kinds of proteins which may be conjugated with polysaccharides such as hyaluronic acid, chondroitin sulfuric acid, and other types of macromolecular polysaccharides.

Spreading of dyes, colloidal particles, or many kinds of biological substances through the ground substance would depend not only on diffusion through homogeneous liquid phases, but also on physicochemical processes at the discontinuous boundaries. These processes would include short range chemical and electrical attractions and other kinds of adsorption processes. The attractions involve not only hydrogen bonding, ionic, dipole-dipole and dipole-induced dipole forces, but also short-range interactions with the aqueous dispersion medium. Thus the spreading process in heterogeneous biological systems depends on the chemical morphology of all the cells and tissues at the various boundaries of discontinuity. These boundaries are subject to many kinds of biological influences: neural and hormonal stimulation, aging, and the effects of various kinds of 'spreading factors' such as the enzyme hyaluronidase (*Joseph et al.*, 1961; *Gersh and Catchpole*, 1960; *Engel et al.*, 1961).

The magnitudes of surface area and net electrical charge have been estimated for the walls of the vacuoles in certain kinds of connective tissue. The values are based on the dimensions of the vacuoles, as observed by *Gersh and Catchpole* (1949, 1960), *Engel et al.* (1961), *Gersh et al.* (1957), *Bondareff* (1957), *Dennis* (1959), and *Chase* (1959). On the basis of such observations,

Engel et al. (1961) calculated the values of certain parameters related to the submicroscopic vacuoles of ground substance.

Submicroscopic dimensions of vacuoles

Diameter of vacuoles	600–1,200 Å	–
Average radius	500 Å	–
Surface area of 1 vacuole, S	3.14×10^{-10} cm^2	$S = 4 \pi r^2$
Volume of 1 vacuole, V	5.25×10^{-16} cm^3	$V = 4/3 \pi r^3$
Vacuoles per cm^3 water-rich phase	1.9×10^{15}	$N = 1/V$
Density of negative electrical charge per 1,000 Å^2		
Dermis	5	–
Cartilage	17	–

From these calculations it appears that irreversible processes of diffusion, spreading, transport and Brownian movement are greatly complicated by the chemical morphology of the colloid-rich phase of the ground substance. These properties would vary with physicochemical state, as it depends on biological processes such as cellular metabolism, uptake of nutrients, exchanges of water, electrolytes, carbon dioxide and oxygen. Perhaps, in certain limiting cases, the principles of statistical mechanics developed by *Boltzmann, Gibbs* and *Einstein* could be usefully applied to systems of biological cells and tissues. However, in these systems, the processes may be found to depend predominantly on chemical morphology, especially as it applies to the distribution of water, electrolytes and other solutes in heterogeneous systems of immiscible colloidal aggregates. Here is is necessary to consider the restrictions on Brownian movement, dispersion and diffusion that were emphasized by *Frey-Wissling* (1953), who does not hesitate to include the possibility of 'purposefulness'. His restrictions apply to all intracellular structures. These must be highly organized to account for well-ordered biological processes. Thus all irreversible processes of transport, spreading and metabolism are to be accounted for by morphological structure and physicochemical state, rather than by 'purposeless' laws of statistical probability and chance, as applied to molecular kinetic theory.

Chapter 5

Biological States

From the point of view of physical chemistry, living cells and tissues differ greatly from the homogeneous mixtures or from the simple suspensions studied in Brownian movement and diffusion. Biological systems which contain cells, extracellular colloidal structures and fluids cannot be treated by the oversimplified methods traditionally applied in classical theories of membrane equilibrium, membrane potentials, osmotic pressure, and ionic transport (*Joseph*, 1971a, b; 1973). Since the work of *Donnan* (1911), it has mainly been assumed that intracellular water is in the liquid state of aggregation. This implies an intracellular solution with perfect mixing of water, electrolytes, proteins and other colloids in a homogeneous mixture constrained only by 'semi-permeable' membranes.

The general conditions for 'equilibrium of heterogeneous substances' developed by *Gibbs* (1875, 1928) were not generally considered. These conditions would require consideration of the chemical potentials and standard chemical potentials of water and inorganic ions in all the heterogeneous phases of living cells and tissues. This requires distinctions between reversible and irreversible processes everywhere in the system. All such processes depend on the states of aggregation of intracellular and extracellular macromolecular substances. These states of aggregation are primarily solid or semi-solid, but they also include water-rich phases which are discontinuous with the highly organized solid phases (*Engel et al.*, 1961).

Discussing the submicroscopic morphology of protoplasm, *Frey-Wyssling* (1953) wrote as follows: 'The substratum in which life is inherent is not a disperse phase with individual particles or ultramicrons; it possesses a *structure*. Its active centers, which control development, are arranged in a given order. They are not intermingled by mere laws of chance and molecular Brownian movement; the fact is rather that they arrange themselves into a delicate, very plastic and flexible pattern, actuated, as it were, by a purposeful coordinated impulse ... For orderly biological processes are unthinkable without presupposing structure, and it is therefore out of the question that any living constituent of protoplasm could consist of structureless, fluid, independently displaceable particles.'

These views as to the submicroscopic structure of protoplasm are in harmony with considerations advanced somewhat earlier by *Schroedinger* (1944). He considered genes and other macromolecular constituents of protoplasm to have the characteristics of anisotropic crystals of very low entropy. The presence of such structures in nucleus and cytoplasm confers a high degree of order to intracellular phases. A certain amount of disorder, however, is necessary to permit the irreversible exchange of metabolites, nutrients, oxygen and carbon dioxide with blood or hemolymph in many organisms; in others the cells may communicate directly with the external environment.

To account for well-ordered or invariant rates of respiratory metabolism in such organisms, *Frey-Wyssling*'s line of reasoning may be applied. Orderly processes of respiratory metabolism are imposed by the structure of the well-ordered intracellular phases of protoplasm. Thus it is not permissible to regard the intracellular phases as being free and unconstrained. It is necessary to treat the distribution of electrolytes and all other soluble substances according to *Gibbs'* equilibrium of heterogeneous substances'. This applies to heterogeneous phases in all possible states of aggregation, and permits the consideration of coexistent states of water in various immiscible or partially miscible phases. Contractility and irritability of many kinds of biological structures involve simultaneous changes of state of water, colloidal aggregates and electrolytes (*Joseph,* 1971a, b; 1973). By changing the solubilities of all reactants, nutrients and complex intermediary products, these biological reactions simultaneously determine the rates of respiratory metabolism. Uniform behavior requires not only well-ordered processes controlled by structures of low entropy, but also irreversible processes of the nature of mixing and 'spreading'. All such processes in the last analysis require labile states of water in miscible or partially miscible phases.

Electromotive Force and Electrolyte Balance

Two main classes of electrolytes may be considered in relation to electrolyte balance in cells and tissues. The first class includes only ions of environmental origin, such as sodium, potassium, calcium, magnesium and chloride. These ions do not react irreversibly in intracellular metabolic processes. They do not enter into reactions within the organism that lead to irreversible change of any kind. In this respect they differ from bicarbonate and hydrogen ions, which are products of intracellular aerobic respiratory metabolism. They must also be distinguished from lactic or pyruvic acid, and other metabolites such as amino acids, creatine or other nitrogenous substances and various kinds of organic phosphates. As a condition of invariance or homeostasis, ions of the first class (Na, K, Ca, Mg and Cl) are transported or exchanged reversibly. Since cells and tissues remain invariant, the *net* rates of transport of these ions are zero. The work of transport, related to the free energy change, is also zero. The condition for reversible transport (homeostasis) is:

$$\Delta G_{rev.} < 0 \tag{5.1}$$

Bicarbonate ions and carbonic acid, produced in intracellular respiration, are transported irreversibly to blood or hemolymph and conducted to the external environment. As a condition for invariance, the rates of transport and respiration must correspond to the rates of formation of CO_2 and heat. At a constant value of the respiratory quotient (RQ), these rates are proportional to the intracellular consumption of oxygen, as determined by the aerobic reactions. The thermodynamic condition for irreversible transport in respiratory metabolism is

$$\Delta G_{irrev.} < 0 \tag{5.2}$$

For a state of invariance or homeostasis, it is further required that m = t, where m is the rate of CO_2 production or oxygen uptake, and where t is the coresponding rate of irreversible transport of one of the gases, the other rate of transport being implied by the RQ. If m ≠ t, the rates of irreversible transport are inadequate to maintain homeostasis, and the cells fall into states of oxygen debt, with accumulation of CO_2, lactic acid and inorganic phosphate.

In addition to being classified as of environmental or of intracellular origin, electrolytes differ with respect to their ionic charges and valence types. The electrolytes of environmental origin fall also into two classes — binary and ternary. The binary class (type AB) is represented mainly by NaCl and KCl. Ternary electrolytes are of type AB_2, as represented by $CaCl_2$ and $MgCl_2$.

In considering the reversible or irreversible transport of the various ions, the following discussion includes univalent and bivalent cations of environmental origin (Na, K, Ca and Mg), the univalent anion of environmental origin (Cl), and the univalent ions of metabolic origin (HCO_3^- and H^+). For ions of the first class (those that are transported reversibly), five equations of reversible electromotive force may be applied (*Gibbs*, 1875, 1928; *Joseph*, 1971a, b; 1973). These are written:

$$
\begin{aligned}
FE + \Delta\mu_{Na} &= 0 \\
FE + \Delta\mu_{K} &= 0 \\
FE + 1/2\, \Delta\mu_{Ca} &= 0 \\
FE + 1/2\, \Delta\mu_{Mg} &= 0 \\
FE - \Delta\mu_{Cl} &= 0
\end{aligned}
\tag{5.3}
$$

where E is the reversible electromotive force, and F is the *Faraday* electrochemical equivalent. The change of chemical potential is denoted as $\Delta\mu$ for each of the ions. These equations express the conditions of reversibility, invariance and stability, which require that isothermal reversible work be zero (*Carnot's* principle). These conditions do not apply to bicarbonate and hydrogen ions, which are transported irreversibly, i.e., with a decrease of free energy. Thus:

$$FE + \Delta\mu_{H^+} > 0$$
$$FE - \Delta\mu_{HCO_3^-} < 0 \tag{5.4}$$

where, as in equations 5.3, $\Delta\mu$ denotes the change of chemical potential of any ion between phase 2 (intracellular) and phase 1 (extracellular). By eliminating FE for the five kinds of ions (equation 5.3),

$$\Delta\mu_{Na} = \Delta\mu_K = \frac{1}{2}\Delta\mu_{Ca} = \frac{1}{2}\Delta\mu_{Mg} = -\Delta\mu_{Cl} = \delta \tag{5.5}$$

where δ is *the equivalent change of chemical potential.* This is expressed as the ratio of $\Delta\mu_i$ to z_i, where z_i denotes the ionic charge, which is taken as negative for chloride or bicarbonate ions.

According to equation 6.4, the condition of irreversible transport of bicarbonate and hydrogen ions requires that

$$\Delta\mu_H > \delta$$
$$\Delta\mu_{HCO_3^-} > \Delta\mu_{Cl} = (-\delta).$$

Then the irreversible transport of H_2CO_3 from metabolizing cells to extracellular phases implies the condition that

$$\Delta\mu_{H^+} + \Delta\mu_{HCO_3^-} > 0$$

and

$$\Delta\mu_{H_2CO_3} > 0.$$

In actively metabolizing cells the chemical potential of carbonic acid is at a higher level than that in extracellular phases. This difference maintains a continuous irreversible flow from intracellular to extracellular phases. Circulation of blood and pulmonary ventilation transport CO_2 irreversibly to air. Thus continuous processes of irreversible transport of the respiratory gases, coordinated with irreversible processes of cellular respiration are necessary conditions for homeostasis. *Carnot's* principle, as applied to reversible work requires the conditions:

$$\Delta G = 0$$

and

$$\Delta H = 0.$$

It may be pointed out in this place that as a condition for physicochemical and physiological invariance, it is also required that for the summation of all intracellular oxidations on a day-to-day basis:

$$\Sigma \; \Delta G = \Sigma \; \Delta H$$

and that

$$\Sigma \; \Delta S = 0.$$

These conditions apply to the coordinated intracellular oxidation of carbohydrates and lipids, and are necessary to account for morphological invariance on a day-to-day basis (*Joseph,* 1971a, 1973).

Free Energy and Chemical Potentials

With the exception of the respiratory gases in the various spaces of lungs, alveoli, bronchioles and those of the upper respiratory system, nose and throat, most tissues of the body are predominantly liquid or solid rather than gaseous. In the terminology of physical chemistry, liquids and solids are condensed phases, in which physicochemical changes involve only small changes of volume, ΔV. Therefore the correlated work terms expressed as $P \, \Delta V$ are also small. In thermodynamics

$$\Delta H = \Delta E + P \, \Delta V$$

and

$$\Delta G = \Delta A + P \, \Delta V.$$

In condensed phases it is possible to make the approximations

$$\Delta H = \Delta E$$

and

$$\Delta G = \Delta A.$$

By the use of the relationship between A and the partition function, f (chapter 3):

$$\frac{\partial A}{\partial m_1} = -RT \left(\frac{\partial \ln f}{\partial m_1} \right) = \mu_1 \qquad (5.6)$$

where m_1 is the mass of the substance 1 (in moles) in a solution in which the masses of all other substances are constant, and where f is the partition function, given by:

$$f = \sum \exp\left(\frac{-\epsilon_1}{kT}\right) \tag{5.7}$$

When all other masses remain constant,

$$\frac{\partial \ln f}{\partial m_1} = -\frac{\epsilon_1}{kT}$$

Therefore

$$\mu_1 = \frac{RT}{k}\epsilon_1 = RT\,N\,\epsilon_1 \tag{5.8}$$

Gibbs' definition of potential is: 'If to any homogeneous mass we suppose an infinitesimal quantity of any substance to be added, the mass remaining homogeneous, and its entropy and volume unchanged, the increase of the energy of the mass divided by the quantity of the substance added is the *potential* for that substance in the mass considered.' The term 'chemical potential' is interchangeable with 'potential', as here defined.

This definition is consonant with equation 5.8, which shows that the potential μ_1 is proportional to the energy of a single molecule of the given substance. When this energy (ϵ_1) is multiplied by *Avogadro*'s number N, we have the molal free energy. This is usually expressed as *Gibbs'* free energy G, but in condensed systems it is practically equal to the *Helmholtz* free energy or work function A.

In general the chemical potential may be related to E and H, as well as to G and A. Thus:

$$\mu_1 = \left(\frac{\partial E}{\partial m_1}\right)_{S,\,V} = \left(\frac{\partial H}{\partial m_1}\right)_{S,\,P} = \left(\frac{\partial A}{\partial m_1}\right)_{T,\,V} = \left(\frac{\partial G}{\partial m_1}\right)_{T,\,P} \tag{5.9}$$

Thus any definition requires the constraints in any given case. Accordingly, equation 5.7 as above is applicable to systems at constant entropy and volume. In mammalian cells and tissues, the constraining conditions are constant temperature and pressure. Then it is possible to define μ in relation to G; in condensed systems it would also be possible to refer it to A (constant temperature and volume). According to equations 5.8 and 5.9, it is possible to relate the chemical potential to thermodynamic functions as well as to the *Boltzmann* probability functions of chapter 3.

Heterogeneous Equilibrium and Electrolyte Balance

In any homogeneous electrolyte solution the chemical potential μ_i of any ion is related to the ionic concentration c_i and the ionic activity coefficient f_i. The notation f_i distinguishes the activity coefficient from the partition function, f, used in the foregoing derivations. In general,

$$\mu_i = \mu_i^0 + RT \ln c_i f_i \tag{5.10}$$

where μ_i^0 is the standard chemical potential. In ordinary aqueous salt solutions f_i is defined by means of a *reference state* at infinite dilution, where its value, by definition, is 1.0. This is impossible in biological systems, which are constrained by conditions of homeostasis or invariance of chemical morphology. These systems never approach states of infinite dilution. In cells and tissues, the variations of μ_i are attributable to variations of μ_i^0 and c_i rather than to variations of f_i. By an appropriate definition of the reference state, f_i may be assigned the value of 1.0 in all phases of a physiological system in its standard state. This is a function of growth, development and aging, which determine chemical morphology in each kind of mammalian cell or tissue (*Joseph*, 1971a, b; 1973). Accordingly, the equation

$$\Delta\mu_i = \Delta\mu_i^0 + RT \ln \frac{c_i''}{c_i'} \tag{5.11}$$

describes the change of chemical potential, $\Delta\mu_i$, the change of standard chemical potential, $\Delta\mu_i^0$, and the concentration ratio c_i''/c_i' between any two phases of a heterogeneous biological system in its standard state. The ion concentration ratios may be estimated from analytical data wherever it is possible to assume complete ionization. In general, this appears to be the case for sodium ions, which are evidently completely ionized in all types of intracellular phases.

Resting electrical potentials, action potentials, and the 'spike' potentials of electrophysiology can be calculated on this basis. It is also possible to estimate $\Delta\mu_i^0$ from the total concentrations of ions of cellular and extracellular phases. The applicable equations are:

$$FE + \Delta\mu_{Na} = 0 \tag{5.12a} (5.3)$$
$$FE^0 + \Delta\mu_{Na}^0 = 0 \tag{5.12b}$$

$$FE_{Na} + RT \ln \frac{c_{Na}''}{c_{Na}'} = 0 \tag{5.12c}$$

where E is the resting potential, E^0 is equal and opposite to the action potential, and E_{Na} is the 'spike' potential. Equations of the same form apply to each of the five physiological ions of environmental origin (equation 5.3). These yield the value of the equivalent change of chemical potential, δ, or the values of $\Delta\mu_i/z_i$ for each of the ions. These are the conditions for isothermal reversible transport, and satisfy *Carnot*'s conditions for invariance and reversibility. The condition for spontaneous irreversible processes with decreases of free energy (negative values of ΔG) are also implied when the formulae are applied to ions such as bicarbonate and hydrogen ions, which are produced in cellular metabolism.

The values of $\Delta\mu_{Na}$ and the other changes of chemical potential may be expressed in either joules or calories. The conversion factors are $1\,J = 43.4\,mV$, and $100\,mV = 2.306\,kcal$. Typical values of E, E^0 and E_{Na} in mammalian skeletal muscle are as follows: $E = -86.8\,mV$ (2 kcal); $E^0 = -130\,mV$ (3 kcal); $E_{Na} = 43.4\,mV$ (1 kcal). For convenience the values are chosen as simple whole numbers, when expressed in kcal. Thus:

$$E = E^0 + E_{Na}$$

and

$$E_a = E - E_{Na}$$

where the action potential is expressed as E_a, which is equal to E^0, but opposite in sign.

Standard Chemical Potential and Dielectric Energy

The change of standard chemical potential of sodium, $\Delta\mu_{Na}{}^0$, is a measure of configurational free energy and entropy of any intracellular phase, and a measure of dielectric constant and dielectric energy. These quantities can be calculated from the total electrolyte concentration of a given phase as referred to the ion concentrations of blood plasma (*Joseph*, 1971a, 1973). The calculations involve consideration of the standard chemical potential of water, $\mu_{H_2O}{}^0$, as estimated from the ion concentrations or freezing point of blood plasma.

The condition for water balance between blood plasma and an intracellular phase is given by the chemical potentials. Thus,

$$\mu_{H_2O}{}' = \mu_{H_2O}{}'' \qquad (5.13)$$

where the potentials refer, respectively, to the extracellular and intracellular phase. This relationship follows from *Gibbs* (1875, 1928), as expressed by his equation 77.

The chemical potential in blood plasma or serum can be estimated from the thermodynamic relation between μ_{H_2O} and the freezing point (*Lewis and Randall*, 1961).

$$\mu H_2 O - \mu H_2 O^0 = 5.26 \, \Delta \tag{5.14}$$

where $\mu_{H_2O}^0$ is the chemical potential of pure liquid water or ice at the freezing point, and where Δ is the freezing point depression, expressed as a negative quantity in centigrade degrees. Thus, when Δ is -0.58 °C, as is typical in mammalian blood plasma, the value of $\mu_{H_2O}{}'$ is about -3.05 cal, as referred to $\mu_{H_2O}^0$. The constant 5.26 cal per degree is the molal entropy of fusion of ice.

The chemical potential of water in plasma and cells may also be estimated from the total ion concentrations in either phase, by means of *Gibbs'* equation 97. Applying this relationship:

$$\Sigma \, m_i{}' \, d\mu_i{}' = \Sigma \, m_i{}'' \, d\mu_i{}'' \tag{5.15}$$

where $m_i{}'$ and $m_i{}''$ denote, respectively, the mass of the ionic species i in plasma and intracellular water, as expressed in moles per kologram water. These values express the free rather than the total ion concentrations.

If, as an approximation, the calculation is based on the extracellular and intracellular concentrations of sodium, potassium and chloride ions, it is found that

$$c_{Na}{}'' \, \Delta\mu_{Na}{}^0 = RT \, [\Sigma \, c_i{}' - \Sigma \, (c_i)''] \tag{5.16}$$

where $\Sigma \, c_i{}'$ refers to the total concentration of Na, K and Cl, which may also be estimated from the freezing point depression. $\Sigma \, (c_i)''$ refers to the total intracellular concentration. The quantity $c_{Na}{}'' \, \Delta\mu_{Na}{}^0$ has been denoted as dielectric energy (*Joseph,* 1971a, b; 1973). The three values of $c_i{}'$ represent the free ion concentrations in blood plasma; Na, K and Cl are assumed to be completely ionized, but the concentrations of ionized potassium and chloride are assumed to be small as compared to the total concentrations, $(c_K)''$ and $(c_{Cl})''$. Then for the two ions:

$$c_K{}'' = \alpha_K{}'' \, (c_K)''$$

and

$$c_{Cl}{}'' = \alpha_{Cl}{}'' \, (c_{Cl})''$$

In adult mammalian skeletal muscle $\alpha_K{}''$ and $\alpha_{Cl}{}''$ are found to be of the order of 0.02 or 0.03. The value of $c_K{}''$ is then of the order of 0.025 mole/kg intracellular water, as compared with similar values for $c_{Cl}{}''$. The value of $\alpha_{Na}{}''$ in all intracellular phases is taken as 1.0, assuming complete ionization. This agrees with the conclusion reached by *Loeb* (1906), who explained his results by assuming very high ionization of Na in intracellular structures in which K was assumed to be mostly in bound forms.

Table I. Electrical potentials and dielectric constant of skeletal muscle[1]

Species	Observed potentials, mV			$\Delta\mu_{Na}^0$ kcal	D''	Reference
	E	E_a	E_{Na}			
Frog	−80	116	36	2.74	30	Nastuk and Hodgkin, 1950
	−92.8	130	37	2.99	28	Nastuk, 1951
	−91.8	125	33	2.85	29	Adrian, 1956
	−80	112	27	2.58	30	Nicholls, 1956
Cat	−79.5	116	36	2.67	30	Trautwein et al., 1953
Guinea pig	−84.5	121	36	2.79	29	Trautwein et al., 1953

[1] From *Joseph* (1973).

In adult skeletal muscle, on the above assumptions, c_{Na}'' is about 0.028 mole/kg water, as compared with a mean value of about 0.140 mole/kg water in blood plasma, a figure that is five times as high. Thus the ratio r_{Na} is about 0.20 on the average. Then, according to equation 5.12c,

$$FE_{Na} + RT \ln r_{Na} = 0.$$

When r_{Na} is taken as 0.20, E_{Na} is −43.4 mV. Then, as explained in the preceding section,

$$E_a = E − 43.4 \text{ mV}.$$

If $\Delta\mu_{Na}^0$ is taken as 3.0 kcal, E_a, the action potential, is found to be 130 mV, and E, the resting potential is −86.8 mV. Converting to heat units, $\Delta\mu_{Na} = 2$ kcal, $\Delta\mu_{Na}^0 = 3$ kcal, and $RT \ln r_{Na} = −1$ kcal (equations 5.12a, b, c). These values are of approximately the same orders of magnitude as those found in electrophysiology for mammalian and frog skeletal muscle (table I). If the value of 2.8 kcal is assumed for $\Delta\mu_{Na}^0$ in mammalian muscle, and if c_{Na}'' is taken as 0.028 mole/kg water, then the dielectric energy is:

$$c_{Na}'' \, \Delta\mu_{Na}^0 = 0.028 \times 2.8 \text{ kcal} = 78.4 \text{ cal}$$

This value estimated from the action potential may be compared with the value calculated from

$$[\Sigma \, c_i' − \Sigma \, (c_i)''],$$

shown in equation 5.16. The value of $\Sigma\, c_i{}'$, as estimated from typical values for mammalian blood serum is 0.315 mole/kg water. The intracellular value of $\Sigma\,(c_i)''$ is taken as 0.185 mole/kg water in human skeletal muscle. Then

$$c_{Na}{}''\,\Delta\mu_{Na}{}^\circ = RT\,(0.315 - 0.185) = 620 \times 0.130 = 80.6\ \text{cal/kg water.}$$

Therefore, the dielectric energy corresponds at 37 °C to configurational free energy and configurational entropy, as they are related to electrolyte and water balance.

The corresponding value of $\Delta\mu_{Na}{}^\circ$, as calculated from electrolyte distribution, is 2.88 kcal, leading to a calculated value of 125 mV for the action potential, E_a, in human skeletal muscle. This may be compared with observed values ranging from 112 to 130 mV in mammalian and frog skeletal muscle (table I).

Dielectric Constant and State of Water

To calculate the standard chemical potentials of such ions as Na, K and Cl from electrostatic theory, we assume that any ion (for example, Na) may be represented as a sphere of radius, r, and an electrical charge, $z_i e$, where e is the electrostatic unit of charge, 4.8×10^{-10} electrostatic units (ESU) (*Born*, 1920). If this charge is uniformly distributed over the surface of a sphere of radius, r, the electrostatic energy may be calculated as the work, w, of charging the sphere from an uncharged condition to the state of an ion of charge $z_i e$. It is assumed that the charge of an ion increases from zero to an integral value $z_i e$ by a gradual continuous process. Then the electrostatic potential of the uniformly charged sphere is calculated on the basis of electrostatic theory to be

$$\psi_i = \frac{z_i e}{D\, r_i}$$

where D is the dielectric constant of the solvent or dispersion medium. On this basis the work, w_i, of charging a single ion is:

$$w_i = \frac{z_i{}^2\, e^2}{2\, D\, r_i}$$

For N ions, the calculation yields a value for the molar work of charging, W_i:

$$W_i = \frac{N\, z_i{}^2\, e^2}{2\, D\, r_i}$$

where N is *Avogadro*'s number and D is the dielectric constant. For a univalent ion such as sodium, this yields:

$$W_i = \frac{6.02 \times 10^{23} \times (4.8 \times 10^{-10})^2}{2\,D\,b_i}$$

where b is expressed in angstrom units. Then

$$W_i = \frac{205\,z_i^2}{D\,b_i}\ \text{kcal.}$$

In the modification of *Born*'s theory introduced by *Laidler and Pegis* (1957), a corrected value of the ionic radius was introduced. This is denoted as b_i and is taken as 1.25 r_i, the radius obtained from crystal structure (*Goldschmidt*, 1926; *Pauling*, 1944). Then in water the product 1.25 D may be taken as 100, and we find:

$$\frac{2.05\,z_i^2}{1.25\,r_i} = \frac{1.64\,z_i^2}{b_i}\ \text{kcal.}$$

For sodium, the crystalline radius is observed to be 0.98 Å. Therefore b_{Na} may be taken as 1.25 Å with a small error. Then for the physiological ions, we obtain the following series of values for W_i, in kcal:

	Ion				
	Cl	K	Na	Ca	Mg
r_i, Å	1.81	1.33	0.98	1.06	0.75
$W_i = \mu_i^0$	0.90	1.23	1.64	6.15	8.20
μ_i^0/μ_{Na}^0	0.55	0.75	1.00	3.75	5.00

When referred to μ_{Na}^0 as a standard, the values are proportional to z_i^2 and inversely proportional to the crystal radius r_i. From the foregoing calculations, an applicable working approximation for $\Delta\mu_i^0$ of any of the five ions may be derived:

$$\Delta\mu_i = \frac{1.64\,z_i^2}{b_i}\left(\frac{1}{D''} - \frac{1}{80}\right) \tag{5.17}$$

where D" is the intracellular or intramuscular dielectric constant, and where the dielectric constant of water (or of blood dialysate) is taken as 80. The corrected ionic radius (1.25 r_i) is denoted as b_i. For a series of values of D″ ranging from 25 to 80, a corresponding series of values for $\mu_{Na}{}^0$ may be obtained:

	D″						
	25	30	40	50	60	70	80
$\mu_{Na}{}^0$, kcal	5.13	4.17	3.28	2.62	2.19	1.88	1.64
$\Delta\mu_{Na}{}^0$, kcal[1]	3.49	2.73	1.64	0.98	0.55	0.24	0.00

[1] Referred to 1.64 kcal.

For mammalian and frog skeletal muscle, the values of the action potential range from 112 to 130 mV (table I). The corresponding values of $\Delta\mu_{Na}{}^0$ are 2.58–2.99 kcal. For an action potential of 130 mV and a change of standard chemical potential of sodium of 2.99 kcal, the value of D″ is found to be 25. Thus values of intracellular dielectric constant of skeletal muscle are generally in the range of 25–35; similar values are found for warm-blooded and cold-blooded vertebrates as well as for invertebrates (*Joseph, 1973*).

Standard Free Energy and Electrolyte Balance

From the results of the preceding section, the values of $\Delta\mu_{Na}{}^0$ and $\Delta\mu_K{}^0$ for any value of the intracellular dielectric constant, D″, may be computed. For skeletal muscle, assuming a value of 30 for D″ for the dielectric constant, the respective values are 2.73 and 2.05 kcal/mole of free ions. Then:

$$\Delta\mu_{Na} = 2.73 + RT \ln r_{Na}$$

and

$$\Delta\mu_K = 2.05 + RT \ln r_K$$

Since the equivalent change of chemical potential, δ, is the same for the two ions, $\Delta\mu_{Na}$ and $\Delta\mu_K$ can be eliminated from the two equations, yielding:

$$\Delta\mu_{Na}{}^0 - \Delta\mu_K{}^0 = RT \ln \frac{r_K}{r_{Na}} \tag{5.18}$$

or

$$(\Delta G^0{}_{K,\,Na}) = -RT \ln \left(\frac{c_K{}'' \times c_{Na}{}'}{c_K{}' \times c_{Na}{}''} \right) \tag{5.19}$$

where the concentrations c_K'' and c_{Na}'' refer to the free ions in the intracellular phase, and where c_{Na}', c_K' refer to the concentration in blood plasma, or other liquid phases of the milieu intérieur. The quantity $(\Delta G^0{}_{K, Na})$ is defined as the *standard free energy of potassium-sodium distribution.* It is equal to the difference of the two changes of standard chemical potential $(\Delta\mu_{Na}{}^0 - \Delta\mu_K{}^0)$. Thus:

$$(\Delta G^0{}_{K, Na}) = 2.73 - 2.05 = 0.68$$

and

$$RT \ln \frac{r_K}{r_{Na}} = 0.68.$$

Converting to common logarithms at the temperature 37 °C,

$$1.419 \log_{10} \frac{r_K}{r_{Na}} = 3.0.$$

In the intramuscular phase of adult mammals, $(c_{Na})''$ is about 28 mM per kg water; assuming sodium to be completely ionized, this is also the value of c_{Na}''. Taking the standard value of c_{Na}' as 140 mM per kg water, the value of r_{Na} is 0.20. Therefore, r_K is estimated as 0.60, yielding a value of 3 mM per kg water for c_K''. This is based on a standard value of 5 mM per kg water for c_K'. In adult mammalian skeletal muscle, typical values of the order of 120 or 130 mM per kg water are usually found. Then the ionization constant for intracellular potassium is denoted as α_K'', and is estimated as

$$\alpha_K'' = \frac{3}{120} = 0.025.$$

For adult mammalian skeletal muscle, typical values of α_K'' are found to be of the order of 0.02–0.03, as compared with values of 1.0 for α_{Na}''. The values of α_{Cl}'' are of the same order of magnitude as those of α_K''. General formulae for the *apparent standard free energies* of distribution of potassium and chloride ions, referred to sodium are expressed as:

$$(\Delta G^0{}_{K, Na})' = - RT \ln \frac{(r_K)}{(r_{Na})} \tag{5.19a}$$

and

$$(\Delta G^0{}_{Na, Cl})' = - RT \ln (r_{Na})(r_{Cl}) \tag{5.19b}$$

where (r_K), (r_{Cl}) and (r_{Na}) refer to the distribution ratios for the total ion concentrations (free plus bound). Similar formulae are applicable to (r_{Ca}) and (r_{Mg}):

$$(\Delta G^0_{1/2Ca, Na})' = -RT \ln \frac{(r_{Ca})^{1/2}}{(r_{Na})} \tag{5.19c}$$

$$(\Delta G^0_{1/2Mg, Na})' = -RT \ln \frac{(r_{Mg})^{1/2}}{(r_{Na})} \tag{5.19d}$$

Thus there are four values of the apparent standard free energies of distribution of K, Ca, Mg and Cl, referred to Na. These represent four thermodynamic equations between five ions. A fifth equation is necessary to determine the five values of $(c_i)''$, as referred to the five values of $(c_i)'$ in an extracellular liquid phase in which the dielectric constant is 80. The fifth equation represents the condition of acid-base balance in the intracellular phase:

$$x + \Sigma (c_i)'' z_i = 0 \tag{5.20}$$

where x denotes the intracellular colloidal charge, expressed as mEq/kg water and where $(c_i)'' z_i$ is also expressed as mEq. The intracellular colloid is anionic, so that x has a negative sign. This leads to an excess of inorganic cations over chloride ions in muscle, which can often be approximated by the expression:

$$x = (c_{Na})'' + (c_K)'' - (c_{Cl})'' \tag{5.21}$$

In muscle, the value of x is of the order of 100–150 mEq/kg water. Thus for the foregoing values of the intracellular ion concentrations:

$$x = 120 + 28 - 20 = 128 \text{ mEq/kg water.}$$

The mean equivalent weight EW of the intracellular macromolecular polyelectrolytes may then be estimated from the formula:

$$EW = \frac{W}{x}$$

where W is the weight of the colloid per kilogram water. This is of the order of about 330 g/kg. Then the mean equivalent weight is estimated as 330/0.128, or 2,570 g/Eq.

Electrolyte balance depends on the standard chemical potentials of the five ions in the two phases. The extracellular values in the liquid phases are standard values for a dielectric constant of 80. Thus there are five equations for $\Delta\mu_i^0$; these depend on the intracellular dielectric constant. Four intracellular values of α_i'' (K, Cl, Ca and Mg) likewise depend on D'' or the intracellular state; the value of α_{Na}'' is taken as 1.0. Electrolyte balance depends also on the charge, x, as related to acid-base balance.

The physicochemical state is described by a minimum of three parameters for each of the five ions, a total of 15 parameters. These include for each ion

$\Delta\mu_i^0$, α_i'' and $(c_i)''$, which are sufficient to establish δ, $\Delta\mu_i$, and E as parameters of state. A mimimum of 15 equations are therefore required to describe the electrolyte balance in the standard state. These include:

(1) Five equations relating μ_i^0 to D''.

(2) Five equations relating α_i'' to D'' or the intracellular state of water.

(3) Four equations relating the five values of $\Delta\mu_i$ (the conditions for stability and invariance, equations 5.5).

In order to obtain these 15 equations, it is necessary to introduce two parameters (D'' and x). The values of the standard chemical potentials, standard free energies of distribution and ionization constants depend on the intracellular states of water and the corresponding dielectric constant. The value of x is determined by the colloidal charge and mean equivalent weight; it depends on colloid-water composition and on the nature of the polyelectrolytes. These are primary properties that are directly related to chemical morphology, as determined by growth and development of the given cell or tissue. Secondary properties of muscle include the action potential, the resting potential, the dielectric energy, c_{Na}'' $\Delta\mu_{Na}^0$, and the rates of respiratory metabolism, as described by irreversible transport of CO_2, O_2 and heat. The list of secondary properties may be extended indefinitely to include maximal work and tension, types of metabolism, and transport of nutrients and metabolites. These depend on changes of state of muscle in biological responses. In large part they depend on changes of state of water related to its properties as a dielectric and as a solvent.

In general, the behavioral responses in contractile tissues such as skeletal muscle and heart depend on simultaneous changes of state of water and electrolytes induced by configurational changes of contractile proteins. In these responses water and electrolytes change from a well-ordered, structured state of low dielectric constant and low entropy to a disordered state of high entropy and lowered free energy. The change of state in muscular contraction is converted to external work, and corresponds to a loss of dielectric energy, and to changes of aerobic and anaerobic respiratory energy. The behavioral responses with respect to external work and respiratory metabolism depend on morphological structure, which has developed through irreversible ontogenetic and phylogenetic processes.

Chapter 6

Biological Development

In any system of cells and tissues the primary function of any structure is to exist. This depends on the phylogenetic and ontogenetic development of chemical morphology. The classification of any system of cells and tissues is the essential first step in determining the laws of its structure and behavior. This requires the determination of the electrolyte composition as it is related to morphological development within any species. As has been shown in the preceding chapter, electrolyte composition of skeletal muscle depends on the dielectric energy, standard chemical potentials of the ions, configurational entropy and free energy. Biological behavior that involves processes of muscular contraction and production of external work then depends in part on chemical morphology, as it develops ontogenetically and phylogenetically.

All processes of growth, development, aging and phylogenetic change involve not only irreversible biological processes of nutrition and chemical synthesis, but also reversible processes of transport and exchange of water and the ions of environmental origin. Development and change therefore take place under many thermodynamic conditions of reversibility, invariance and constraint. These conditions establish a principle which has been called the 'conservation of reversibility' (*Joseph, 1973*). In mammals, for example, the conditions of a constant milieu intérieur have led to phylogenetic invariance of the blood electrolytes extending over entire geological eras (*Bernard, 1878, Macallum, 1910, 1926*). Certain properties relating to the physicochemical nature of water and the inorganic electrolytes are constant, not only in the milieu intérieur of vertebrates and invertebrates, but also in the well-ordered solid or semi-solid phases, including skeletal muscle, heart, brain and nervous system. The invariance depends on the constancy of the chemical potentials of water and electrolytes of any physiological system in all stages of its biological development. Phase rule conditions of invariance and constraint depend on the reversible transport of water, sodium, potassium and chloride in all types of cells and tissues.

Irritability, contractility and responsiveness of tissues such as skeletal muscle and heart depend on the lability of water, electrolytes and macromolecular polyelectrolytes. Chemical morphology of any biological structure is a *primary* function, which determines all the *secondary* properties. The latter set of properties or attributes includes intracellular respiratory metabolism, involving irreversible production of CO_2 and heat, as well as uptake of nutrients, metabolites and oxygen. These are well-ordered biological processes which necessarily imply well-ordered morphological structures at cellular and submicroscopic levels (*Frey-Wyssling*, 1953). Reversible work in an isothermal invariant system is zero (*Carnot*'s principle).

Thus the standard resting electrical potential or electromotive force is a constant secondary property that depends on chemical morphology. Irreversible work, as in muscular contraction, depends on changes of state that imply release of dielectric energy, configurational free energy, and chemical or electrical potentials. These are related to changes of state of the intramuscular water, which in turn leads to changes of metabolic rates and types of metabolism. Maintenance of a state of high configurational free energy and low configurational entropy requires nutrient supplies of 'negative entropy'. The principal source of this is intracellular or extracellular glucose or other carbohydrates, which supply negative entropy, and which are therefore oxidized with increases of entropy. The responses of respiratory metabolism are then induced by the configurational changes of state of water in behavioral responses of muscle. If we denote a set of morphological parameters as M, and a set of metabolic parameters as m, then

$$M \supset m, \text{ and } M' \supset m'.$$

Thus morphology implies metabolism, and a modified state, M', implies a metabolic response, m'. Here, M would refer, for example, to uncontracted muscle fibers, and M' would refer to the contracted state. Thus the change of state involving both M and m is implied by the particular biological process, P, which describes the behavioral response. In symbolic notation,

$$P \supset (M, m).$$

In behavioral responses, morphological and metabolic processes are interdependent rather than independent. Freedom and independence are to a large extent eliminated by constraints that are imposed by chemical morphology and thermodynamic conditions of balance that apply to all the internal reversible processes. The second law of thermodynamics (*Carnot*'s principle) can be applied in the mathematical form of conservative line integrals (*Caratheodory*, 1909; *Joseph*, 1973). In this form, it may be stated that there are reversible accessible states in the neighborhood of any inaccessible or relatively inaccessible state. Thus biological behavior is channeled into certain states of normal morphological and metabolic responses. Conforming to the well-ordered behavioral responses or processes, there are concurrent morphological and metabolic changes which depend on the ontogenetic and phylogenetic development of each kind of structure.

Chemical Morphology of Oocytes

Electrolyte composition of the eggs of the sea urchin *Paracentrotus lividus* has been studied by *Rothschild and Barnes* (1953) with the following results:

	Eggs	Sea water	Ratio (r_i)	($\Delta G^0_{i, Na}$)
Na	52	485	0.11	–
K	210	10	21	−3.18
Ca	4	11	0.36	−0.98
Mg	11	55	0.20	−0.84
Cl	80	560	0.14	2.55

The values are expressed as millimoles per kilogram water; the ratios of the cellular concentrations to those of sea water are shown in the third column of figures. Calculated values of the apparent standard free energies of ion distribution referred to sodium are shown in the fourth column. These are obtained from the analytical figures by means of equations 5.19, a, b, c, d. At temperatures of about 20–30 °C, the following approximation is sufficiently accurate:

$$(\Delta G^0_{i, Na})' = -1.4 \log \frac{(r_i)^{1/z_i}}{(r_{Na})}$$

Thus, for potassium:

$$(\Delta G^0_{K, Na})' = -1.4 \log \frac{21}{0.11} = -3.18 \text{ kcal.}$$

For chloride

$$(\Delta G^0_{Na, Cl})' = -1.4 \log (0.11)(0.14) = 2.55 \text{ kcal.}$$

For the bivalent cations, the numerator is expressed as $(r_{Ca})^{1/2}$ or $(r_{Mg})^{1/2}$. The four respective values of $(\Delta G^0_{i, Na})'$ will be shown to be comparable in magnitude to corresponding values of skeletal muscle of invertebrates, cold-blooded vertebrates and mammals.

In the case of $(\Delta G^0_{K, Na})'$, it is also possible to compare the results with those obtained for oocytes of the frog *Rana pipiens* (*Naora et al.*, 1962). Expressing the concentrations as mEq/kg water, they found for cytoplasma and nucleus:

	Na	K
Cytoplasm (C)	60	106
Nucleus (N)	227	258
N:C ratio	3.8	2.1
(r_i) (cytoplasm)	0.59	42
(r_i) (nucleus)	2.23	103

The extracellular ion concentrations refer to frog blood serum or to frog Ringer solution. The values of $(\Delta G^0_{K, Na})'$ are estimated as follows:

K, Na, cytoplasm = -2.60 kcal
K, Na, nucleus = -2.16 kcal

These values for cytoplasm and nucleus of frog oocytes are positive with respect to the value -3.18 kcal estimated for sea urchin eggs. The value -2.60 kcal found for the cytoplasm of frog eggs is more nearly comparable to that for the invertebrate species than is the value -2.16 kcal found for the nucleus of the amphibian eggs. The value -2.60 kcal is quite similar to values found for $(\Delta G^0_{K, Na})'$ in many types of adult skeletal muscle.

Chemical Morphology of Skeletal Muscle

Values of the apparent standard free energies of K, Na distribution for the skeletal muscle of a marine invertebrate are computed from the results of *Robertson* (1961). For the whole muscle of the lobster *(Nephrops norvegicus)* the results yield:

	Muscle m M per kg H_2O	Blood serum m M per kg H_2O	(r_i)	$(\Delta G^0_{i, Na})'$ kcal
Na	83.2	577	0.16	–
K	166.5	8.6	19.7	-2.82
Ca	5.2	16.2	0.32	-0.71
Mg	19.1	10.4	1.85	-1.28
Cl	109.9	527	0.21	2.03

These values may be compared with those for the frog *(Katz, 1895; Fenn, 1935).*

	Muscle mEq/kg H_2O	Blood serum mEq/kg H_2O	(r_i)	$(\Delta G^0_{i, Na})'$ kcal
Na	29.3	105.8	0.28	–
K	96.7	2.5	38.7	-2.92
Ca	4.4	3.0	1.47	-0.89
Mg	11.8	2.0	5.9	-1.26
Cl	5.9	74.3	0.18	1.73

Thus the four estimated values of $(\Delta G^0_{i, Na})'$ are of quite similar magnitudes for the skeletal muscle of the fresh water amphibian, as compared with the values for the marine invertebrate muscle. The values for potassium-sodium distribution for the two species (-2.82 and -2.92 kcal) are similar to the value obtained for sea urchin eggs (-3.18 kcal) Thus with respect to potassium-sodium distribution, the protoplasm of oocytes shows similar properties to those of skeletal muscle of cold-blooded species.

An extensive comparison of the apparent standard free energies of distribution of K, Ca, Mg and Cl, as referred to sodium is presented in table II. This is based on calculations for skeletal muscle of mammals, cold-blooded vertebrates and invertebrates (*Joseph*, 1973).

The results of table II show considerable regularities when they are compared with respect to the differences between invertebrates and either warm-blooded or cold-blooded vertebrates. With the exception of the intracellular phase of lobster muscle, the values of $(\Delta G^0_{K, Na})'$ fall within the range of -2.61 kcal (pecten) to -3.19 kcal for the parietal muscle of the eel. For four species of mammalian muscle, the values vary between narrow limits (-3.02 to -3.09 kcal). There are similar regularities when the various values of $(\Delta G^0_{1/2\ Ca, Na})'$ and $(\Delta G^0_{1/2\ Mg, Na})'$ are compared with respect to the various vertebrate and invertebrate species. Thus in mammals, the values of $(\Delta G^0_{1/2\ Mg, Na})'$ fall within the range -1.71 to -1.77 kcal as compared with values of from -0.83 to -1.03 kcal for $(\Delta G^0_{1/2\ Ca, Na})'$.

Perhaps the most noteworthy difference among the various species occurs when they are compared with respect to the values of $(\Delta G^0_{Na, Cl})'$. For mammals, the values for sodium chloride distribution range only from 1.95 to 2.10 kcal. Values for the cold-blooded vertebrates are not greatly different (1.61 kcal for the eel to 2.01 kcal for salmon muscle). However, the values for invertebrate muscle range from 2.03 kcal (lobster) to 2.73 kcal *(Mytilus edulis)*. The value for the intracellular phase of lobster muscle is 3.19 kcal. These species are marine invertebrates; the muscles are exposed to extracellular fluids of high electrolyte concentrations. In this type of organism, the freezing point depression may be of the order of 1.6–2.0 °C and the corresponding values of the chemical potential of water (μ_{H_2O}) may be as low as -10.5 cal as referred to the potential of water. In this situation, osmoregulation is managed by the production of various kinds of nitrogenous substances such as amino acids and trimethylamine oxide.

Table II. Apparent standard free energies of ion distribution (values in kcal/Eq)

Species	Skeletal muscle			
	K, Na	$\frac{1}{2}$ Ca, Na	$\frac{1}{2}$ Mg, Na	Na, Cl
Mammals				
Man	-3.02	-0.84	-1.71	1.95
Pig	-3.05	-0.83	-1.77	2.10
Rabbit	-3.09	-1.03	-1.77	1.95
Rat	-3.05	-0.98	-1.75	2.03
Cold-blooded vertebrates				
Frog	-2.92	-0.89	-1.26	1.73
Salmon	-3.02	$-$	$-$	2.01
Eel				
Parietal	-3.19	-0.43	-0.89	2.00
Tongue	-2.98	-1.03	-1.44	1.61
Invertebrates				
Carcinus maenas	-2.69	-1.05	-1.24	2.69
Pecten	-2.61	$-$	$-$	2.70
Mytilus edulis	-2.62	$-$	$-$	2.73
Lobster				
Whole muscle	-2.82	-0.71	-1.26	2.03
Intracellular	-3.65	-1.38	-1.61	3.19
Oocytes				
Sea urchin	-3.18	-0.98	-0.84	2.55
Human lens	-3.04	-0.89	-1.33	1.61
Human erythrocytes	-3.20	$-$	$-$	1.82

It is clear that the standard free energies of ion distribution are universal and *non-specific* thermodynamic functions that cannot be directly related to any *secondary* functions such as respiratory metabolism, permeability or *irreversible* transport of ions. The standard free energies are characteristic of nearly all reversible processes of ion exchange between intracellular phases of protoplasm and extracellular phases of blood plasma, hemolymph, or the *milieu intérieur* or *extérieur*. All such reversible processes are primary and independent of secondary irreversible processes of respiratory metabolism or transport.
Values computed in *Joseph* (1973).

In the salt water habitat of the sea urchin, the chloride concentration is 560 mM per kg water. Comparable concentrations are found in the blood or hemolymph of lobsters, crabs and in the extracellular fluids of other marine invertebrates. In these species the presence of various nitrogenous osmoregulatory substances lowers the intracellular chemical potential of water to that of the extracellular fluid. This enables the tissue to maintain relatively low concentrations of both sodium and chloride. Thus in sea urchin eggs (r_{Na}) is 0.11 and (r_{Cl}) is 0.14. Comparably low values are found in the muscles of marine invertebrates. Low values of (r_{Na}) and (r_{Cl}) are responsible for values of the order of 2.5–3.2 kcal for NaCl distribution in the protoplasm of sea urchin eggs and the intracellular phase of lobster muscle. The high values are due to osmoregulatory mechanisms that maintain intracellular sodium and chloride concentrations at physiological levels.

Development of Skeletal Muscle

Apparent standard free energies of distribution of K, Ca, Mg and Cl as referred to Na have been estimated for human and pig skeletal muscle in three states of development: fetal, newborn and adult. The values of $(\Delta G^0_{i, Na})'$ for the various ions have been calculated from analytical results (*Dickerson and Widdowson,* 1960; *Widdowson and Dickerson,* 1964). The calculated results are presented in table III.

Table III. Apparent standard free energies in developing skeletal muscle (in kcal/Eq)

	Fetus	Newborn	Adult
K, Na			
Man	−1.28	−2.05	−3.02
Pig	−1.53	−2.75	−3.05
½ Ca, Na			
Man	0.00	−0.43	−0.84
Pig	−0.23	−0.90	−0.83
½ Mg, Na			
Man	−0.62	−1.73	−1.71
Pig	−0.53	−1.87	−1.77
Na, Cl			
Man	0.28	0.49	2.10
Pig	0.28	1.53	2.10

It is evident that the results for all values of the apparent standard free energies of ion distribution tend to agree in both species in the adult as well as in the fetal state. Only in the case of $(\Delta G^0_{K, Na})'$ are there fetal values that do not tend to approach zero. These values are -1.28 kcal (man) and -1.53 kcal (pig). This indicates appreciable binding of potassium even in the fetal state, where $\alpha_K{''}$ is about 0.063 (man) and 0.036 (pig) (*Joseph*, 1973). The fetal values of $\alpha_{Cl}{''}$ are 0.60 (man) and 0.17 (pig). In the adult state $\alpha_K{''}$ is 0.024 (man and pig), while $\alpha_{Cl}{''}$ is 0.019 (man) and 0.011 (pig). With respect to the ionization constants of K and Cl, the two kinds of muscle show differences during the fetal and postpartum growth periods. At all stages of development during these periods, potassium is more highly bound than chloride. However, in the adult period of man, $\alpha_K{''}$ is similar to $\alpha_{Cl}{''}$; in the pig $\alpha_{Cl}{''}$ is considerably lower. In both species the values of α change considerably during growth and development, but potassium ion binding is less dependent on age than that of chloride. The results show that binding of potassium changes with the synthesis of the cellular colloids (*Loeb*, 1906). This occurs in plants as well as in animal protoplasm. Analogous binding of cations and anions to various kinds of synthetic macrocyclic compounds has been established (*Christensen et al.*, 1971).

The values of $(\Delta G^0_{i, Na})'$ are quantitative measures of the deviations of ion distributions from the classical 'membrane equilibrium' of *Donnan* (1911). When two aqueous solutions of dielectric constant 80 are separated by a semi-permeable membrane which permits only the passage of water and the dialyzable electrolytes, the values of $\Delta\mu_i^0$ become zero for all the ions. In the absence of chemical binding to the nondialyzable proteins or polyelectrolytes, the values of the ionization constants become 1.0 for each kind of ion. Then the conditions of equilibrium between the homogeneous inner colloidal solution and the outer dialysate require that:

$$\Delta\mu_{Na} = \Delta\mu_K = \frac{1}{2}\Delta\mu_{Ca} = \frac{1}{2}\Delta\mu_{Mg} = -\Delta\mu_{Cl} = \delta \qquad (6.1)$$

where δ is the equivalent change of chemical potential. In general, for any of the dialyzable ions in the two phase system:

$$\Delta\mu_i = \Delta\mu_i^0 + \frac{RT}{z_i}\ln\frac{(c_i)''}{(c_i)'} + \frac{RT}{z_i}\ln\frac{\alpha_i''}{\alpha_i'} \qquad (6.2)$$

Equations of this form apply to Na, K, Ca, Mg and Cl in systems of living cells and tissues. All the solid or semi-solid colloidal aggregates of protoplasm or the well-ordered structures of connective tissue are heterogeneous rather than homogeneous. In these structures, water is organized in immiscible or partially immiscible phases, and may have solvent or dielectric properties much different from those in the liquid state of dielectric constant 80.

Under the conditions of the classical 'membrane equilibrium', the terms $\Delta\mu_i{}^0$ and $RT \ln \alpha_i{}''/\alpha_i{}'$ become zero. Then for each of the five ions:

$$\Delta\mu_i = \frac{RT}{z_i} \ln \frac{c_i{}''}{c_i{}'} = RT \ln r_i{}^{1/z_i} \tag{6.3}$$

Here $c_i{}''$ and $c_i{}'$ refer only to the completely dissociated ions, and r_i refers to the ratio of the free ions. Under these conditions the equilibrium between the two homogeneous solutions of dielectric constant 80 is described by:

$$r_{Na} = r_K = r_{Ca}{}^{1/2} = r_{Mg}{}^{1/2} = r_{Cl}{}^{-1} \tag{6.4}$$

In actual heterogeneous systems of well-ordered structured cells and tissues:

$$\Delta\mu_i = (\Delta\mu_i{}^0)' + RT \ln (r_i) \tag{6.5}$$

where (r_i) refers to the ratio of total ionic concentrations $(c_i)''$ and $(c_i)'$. These values include both the ionized and bound fractions. Then, since $\Delta\mu_{Na}$ and $\Delta\mu_K$ are equal (equation 6.1), we find:

$$(\Delta G^0{}_{K, Na})' = (\Delta\mu_K{}^0)' - \Delta\mu_{Na}{}^0 \tag{6.6}$$

This is based on the assumption that sodium is completely ionized in intracellular and extracellular phases ($\alpha_{Na} = 1.0$). Similar equations apply to the distributions of the five ions. Therefore values of $(\Delta G^0{}_{i, Na})'$ that approach zero may signify that the conditions of the 'Donnan equilibrium' are being approached, i.e., r_{Na} approaches r_K, and $(\Delta\mu_K{}^0)'$ approaches $\Delta\mu_{Na}{}^0$. Values of $(\Delta G^0{}_{i, Na})'$, on the other hand, that deviate greatly from zero, imply coexistence of water in immiscible or partially immiscible states, in which dielectric constants may vary widely. This implies large variations of $\mu_{Na}{}^0, \mu_K{}^0$ and the other standard chemical potentials. It implies variations of $\alpha_i{}''$, which depend on the states of water in cells and extracellular connective tissues.

Dielectric Constant and Dielectric Energy

The standard chemical potential of each of the inorganic physiological ions is related to the dielectric constant D of the solvent or dispersion medium and to the charge and radius of any ion. Thus, as an approximation,

$$\mu_i{}^0 = \frac{N z_i{}^2 e^2}{2 D r_i}$$

where N is *Avogadro*'s number, z_i is the charge, e the electrostatic unit of charge, and r_i is the radius of the ion in the crystalline state. This is a constant for each kind of ion, and is independent of the nature of the other ion with which it forms a crystal (of a binary or ternary type of electrolyte) (*Born*, 1920). In actual practice it has been found necessary to use a corrected value of the ionic radius. This is denoted as b: it is obtained by multiplying r_i by the factor 1.25. Then for sodium ion, $r_{Na} = 0.98$ Å; for b_{Na} we may adopt the value 1.25 Å. The following result is then obtained:

$$\mu_{Na}{}^0 = \frac{N\,e^2}{2\,D\,b_{Na}}$$

In water (D = 80, $b_{Na} = 1.25$ Å):

$$\mu_{Na}{}^0 = \frac{2.05}{b_{Na}} = \frac{2.05}{1.25\,r_{Na}} = 1.64 \text{ kcal.}$$

Then for any ion (Na, K, Ca, Mg or Cl):

$$\mu_i{}^0 = 1.64\,z_i{}^2\,\frac{r_{Na}}{r_i}$$

This leads to the following series of values (in kcal) of $\mu_i{}^0$ in water:

Cl	K	Na	Ca	Mg
0.90	1.23	1.64	6.15	8.20

The theoretical equation for $\Delta\mu_i{}^0$ of each of the five ions is then:

$$\Delta\mu_i{}^0 = \frac{1.64\,z_i{}^2}{b_i}\left(\frac{1}{D''} - \frac{1}{80}\right) \qquad\qquad (6.7)\ (5.17)$$

where D'' denotes the value of the dielectric constant in a phase in which water is in a structured 'ice-like' state. This would refer to intracellular phases in which the water is immiscible or partially miscible with liquid water of dielectric constant 80. It also refers to extracellular phases such as the ground substance of connective tissue, in which the water is well-ordered, and in which D'' may differ from 80, the limiting value in the liquid state of aggregation.

Equation 6.7 may also be written in the form:

$$\Delta\mu_{Na}{}^0 = \mu_{Na}{}^0 - 1.64 \text{ (kcal)}$$

where $\Delta\mu_{Na}{}^0$ is the value of the standard chemical potential in a phase in which the dielectric constant is D''. Then the value of D'' may be estimated from the formula:

$$D'' = \frac{131}{1.64 + \Delta\mu_{Na}{}^0} \tag{6.8}$$

In this formula the constant 131 is the product of (1.64×80): this yields the value of 80 for D'' when $\Delta\mu_{Na}{}^0$ is zero (classical *Donnan* equilibrium). When $\Delta\mu_{Na}{}^0 = 1.64$ kcal, the value of $\mu_{Na}{}^0$ is 3.28 kcal and the calculated value of D'' is 40. As was shown in chapter 5, the action potential of skeletal muscle is proportional to $\Delta\mu_{Na}{}^0$. Thus:

$$E_a = 43.4 \; \Delta\mu_{Na}{}^0 \tag{6.9}$$

where the change of standard chemical potential is measured in kcal and E_a is in millivolts. Thus a value of 3 kcal for $\Delta\mu_{Na}{}^0$ corresponds to a value of the action potential of 130.2 mV. From this the value of D'' may be calculated according to equation 6.8:

$$D'' = \frac{131}{4.64} = 28.$$

The value of the intracellular standard chemical potential is a measure of configurational free energy or configurational entropy in the system. It is also a measure of a quantity defined as 'dielectric energy' (*Joseph*, 1971a, b, 1973). This is defined as the product $c_{Na}{}'' \; \Delta\mu_{Na}{}^0$. Thus:

$$c_{Na}{}'' \; \Delta\mu_{Na}{}^0 = RT \left[\Sigma \; c_i{}' - \Sigma \; (c_i)'' \right] \tag{6.10}$$

In adult intramuscular phases of mammals the sodium concentration is of the order of 0.03 mole/kg water, and $\Delta\mu_{Na}{}^0$ is of the order of 2.8 kcal. This yields a value of 84 cal/kg water for the dielectric energy of muscle. Taking the value of 0.315 for $\Sigma \; c_i{}'$ in moles per kilogram water, the calculated value of $\Sigma \; (c_i)''$ is 0.180 mole/kg water. This corresponds well with the experimentally observed values found in human or pig skeletal muscle.

Values of the dielectric energy of many species of vertebrates and invertebrate muscle, as calculated either from the action potential or from the total electrolyte composition $\Sigma \; (c_i)''$ are of the order of 80 kcal/kg water.

In converting to work units (joules) the conversion factors are 1 cal = 4.183 J, and 1 J = 9.8 kg m. Therefore 1 kcal = 0.427 kg m. A value of 80 cal/kg water for the dielectric energy of skeletal muscle therefore corresponds to 335 J/kg water. The biceps and brachialis muscles of an adult man contain a total of about 250 g of water (0.25 kg). Therefore the value of the dielectric energy is about 20 cal or 83.6 J (8.5 kg m). *Hill* (1944, 1951) has estimated that these muscles can deliver about 9 kg m as kinetic energy to an external object such as a cricket ball (150 g). For example, a ball weighing 300 g can be lifted to a height of about 30 m. This represents a potential energy of 9 kg m.

The value is slightly greater than that estimated in the foregoing for dielectric energy (8.5 kg m for the arm muscles, or about 34 kg m per kg water).

If we take a value of 20 cal as the dielectric energy of the human arm muscles, this corresponds to 83.6 J. The weight of a baseball, like that of the cricket ball is about 150 g. The kinetic energy of a thrown ball is therefore:

$$E_{kin} = \frac{1}{2} m v^2 = 75 \, v^2 \text{ (joules)}$$

where v is the velocity. Hence:

$$75 \, v^2 = 83.6 \times 10^7$$

where 83.6 J is the kinetic energy, and 10^7 is the number of ergs in 1 J. The calculated value of v is then 33.3 m/sec, or about 1.2 miles/min. This energy then corresponds fairly well with the maximum that can be delivered by a highly trained pitcher in professional baseball. It corresponds to a calculated value of 80 cal/kg water for the dielectric energy of human skeletal muscle.

A value may also be estimated from the action potential, E_a. This is proportional to $\Delta\mu_{Na}^0$; the conversion factor is 100 mV, equals 2.306 kcal, or 1 kcal = 43.4 mV. Therefore 120 mV corresponds to 2.77 kcal. Then when c_{Na}'' is 0.028 mole/kg water,

$$c_{Na}'' \, \Delta\mu_{Na}^0 = 0.028 \times 2.77 = 77.6 \text{ cal/kg water.}$$

Then, in general,

$$\begin{aligned} \text{Dielectric energy} \quad &= 2.306 \, c_{Na}'' \, E_a \\ &= RT \left[\Sigma \, c_i' - \Sigma \, (c_i)'' \right] \end{aligned} \qquad (6.11)$$

Thus the distribution of ions and the action potential, E_a, are determined by the dielectric energy of the intracellular phase. This depends on the dielectric constant D'' which is related to the state of aggregation of the intracellular water. Thus the following parameters are related to chemical morphology, which is the primary function that implies the value of D'':

(1) Standard chemical potentials

(2) Action potential, E_a

(3) Resting potential, E, which is proportional to $\Delta\mu_i$ of any ion but opposite in sign

(4) Dielectric energy

(5) Electrolyte distribution, as expressed by $\Sigma \, (c_i)''$

(6) Configurational free energy, proportional to dielectric energy, to maximal work and to isometric tension

Since they depend on chemical morphology, these properties are at the same time related to phylogenetic and ontogenetic development. This is indi-

Table IV. Dielectric properties in relation to age and species

Species	$\Delta\mu_{Na}{}^0$ kcal	E_a mV	D''	$c_{Na}'' \Delta\mu_{Na}{}^0$	Reference
Man					
Fetus	0.34	15	66	36.58	Widdowson and Dickerson, 1964
Newborn	0.93	40	51	71.30	
Adult	2.81	122	30	80.60	
Pig					
Fetus	0.37	16	65	43.40	Widdowson and Dickerson, 1964
Newborn	1.09	47	48	71.92	
Adult	3.12	135	28	83.70	
Rat					
2 days	0.37	16	65	41.54	Hazlewood and Nichols, 1969
603 days	2.72	118	30	75.07	Lowry and Hastings, 1952
988 days	1.93	84	37	78.10	Lowry and Hastings, 1952
Frog	2.85	125	29	82.65	Katz, 1895
C. maenas	2.68	116	30	145.3	Shaw, 1955, 1958a, b

Values of $\Delta\mu_{Na}{}^0$, E_a and D'' from table I.

cated by the results of table IV which compares some of the dielectric properties of muscle in relation to age in mammals and in relation to phylogenetic change in adult animals.

From the results of table IV, it is evident that the value of $\Delta\mu_{Na}{}^0$ is approximately 2.7–3.1 kcal, not only in mammalian skeletal muscle, but in such cold-blooded species as the frog and the marine crab, *Carcinus maenas.* This value corresponds to values of the intramuscular dielectric constant D'' that are of the order of 30. This value tends to approach a lower limit for many kinds of intramuscular protoplasm. It corresponds to calculated action potentials of the order of 115–135 mV in adult mammals and in cold-blooded animals such as the frog, which lives in fresh water. Similar values are calculated for marine species such as *C. maenas* (fig. 3).

The values of the dielectric energy of adult mammals such as man, pig and the rat are also fairly constant, and of the order of 80 cal/kg water. For example, the calculated values for the rat are 75.07 cal at 603 days and 78.10 cal at 988 days. The values for man (80.6 cal) and pig (83.7 cal) are of similar magnitudes. A value of 82.65 cal for the frog is likewise similar.

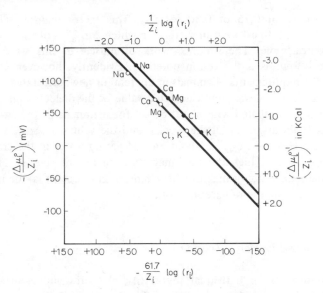

Fig. 3. Distribution of ions in skeletal muscle of poikilotherms, i.e. frog (•) and marine crab, *Carcinus maenas* (○). Ordinates: left: values of apparent changes of equivalent standard chemical potentials, in millivolts; right: values in kcal. Upper abscissa: ion distribution ratios (r_i) on logarithmic scales, lower abscissa: distribution ratios converted to electromotive force, uncorrected for configurational free energies or potentials. Values of (r_i) for *C. maenas* displaced in negative direction (lower intracellular ratios) because of effect of nitrogenous osmoregulatory substances such as trimethylamine oxide (*Joseph,* 1973).

Since $\Delta\mu_{Na}$ is of the same order of magnitude for these species, the intramuscular values of c_{Na}'' are also found to be constant. Thus if $\Delta\mu_{Na}^0$ is assumed to be 3.0 kcal and if $c_{Na}'' \Delta\mu_{Na}''$ is assumed to be 80 cal/kg water, the corresponding value of c_{Na}'' would be about 0.027 mole/kg water. Values of this order of magnitude are found for man, pig, rat and frog in the adult tissues. These values do not apply to the skeletal muscle of newborn animals of the various mammalian species. In man, pig and the 2-day-old rat, the fetal values of D'' are about 65. The values of the dielectric energy are of the order of 40 cal/kg water in the very early stages of life. The calculated action potential, E_a, is also very low in this period.

As explained in a previous section, in the fetal or newborn period, low values of $(\Delta G^0_{i, Na})'$ are found for calcium, magnesium and chloride distribution as referred to that of sodium. These low deviations correspond to states of water that tend to approach the liquid state of aggregation in which D'' is 80. This value is not actually found in any type of muscle tissue, but there is a tendency for it to be approached in the fetal state, where values of about 65 are found. In this state values of about 100 mM per kg water or higher are found for c_{Na}'', corresponding to values of 0.75 or 0.80 for r_{Na}. This factor maintains the dielectric energy at about 40 cal even in systems in which $\Delta\mu_{Na}{}^0$ is of the order of less than 0.40 kcal/mole. The maximal work of muscle in this state of development is very low, as is the action potential. Evidently, however, it is sufficient to meet the requirements of animals in the fetal or newborn states.

In the marine crab, *C. maenas,* the calculated value of the dielectric energy is very high (145 cal as compared with about 80 cal for mammals or the frog). Dielectric constant is about 30 for all the species, and the values of $\Delta\mu_{Na}{}^0$ are also about the same. However, $c_{Na}{}''$ is about 54 mM per kg water in the crab. The dielectric energy is of a high order of magnitude in marine species. This enables them to maintain states of high isometric tension in muscles that are not primarily adapted to production of external work.

Dielectric Energy and Isometric Tension

In skeletal muscle, as in individual myofibrils, elasticity and isometric tension are related to the dielectric energy. Expressing the energy E as a function of entropy and length:

$$dE = T\,dS + t\,dl$$

where t is the tension in grams per square centimeter, and where l is the length in centimeter. Then the free energy of the fiber is related to T and t in the following way:

$$dG = m_{H_2O}\,d\mu_{H_2O} + m_i\,d\mu_i$$

Accordingly, if the muscle is stretched under conditions of constant temperature and constant chemical potential of water, we find

$$dG = -m_i\,d\mu_i = -l\,dt$$

If Na is assumed to be completely ionized, and if the other ions (K, Ca, Mg and Cl) are only slightly ionized, then $c_{Na}{}''$ is of a higher order of magnitude than $(c_i)''$ for the other ions. Also, if the intrafibrillar concentration $(c_{Na})''$ remains constant during a short period of isometric concentration, l may also be treated as a constant. Then, when the tension t is measured as an increase referred to the resting tension,

$$-l\,dt = c_{Na}{}''\,d\mu_{Na}{}^0$$

where $c_{Na}{}''$ is the number of moles per kilogram water, and where l is the length of fiber required to yield 1 kg water. The tension t is measured in grams per square centimeter of cross-sectional area. Then:

$$l = \frac{1,000}{A \, \bar{d} \, w}$$

where $\bar{d}$ is the density of the muscle, and where w is the water content in grams per kilogram muscle. Then when A is 1 cm^2,

$$-t = 10.20 \, \bar{d} \, w \, c_{Na}{}'' \, \Delta\mu_{Na}{}^0$$

where 10.20 is the factor necessary to convert kilogram-meters to joules. If $\bar{d}$ is taken as 1.05, and if w is assumed to be 0.80, then

$$-t = 8.60 \, c_{Na}{}'' \, \Delta\mu_{Na}{}^0 \qquad\qquad (6.12)$$

In isometric contraction produced by tetanic stimulation, $c_{Na}{}'' \, \Delta\mu_{Na}{}^0$ is negative. Muscle passes from a state of low dielectric constant ($D'' = 30$) to a state in which the water becomes disordered, and in which D'' may approach 80. Then, as in frog muscle,

$$\Delta\mu_{Na}{}^0 = 2.85 \text{ kcal and } c_{Na}{}'' \, \Delta\mu_{Na}{}^0 = 82.65 \text{ kcal}$$
$$t - 8.6 \times 0.029 \times 11,920 = 2,970 \text{ g/cm}^2.$$

When $\Delta\mu_{Na}{}^0$ is converted to the action potential, E_a, 1 kcal corresponds to 43.4 mV. Then

$$t = 830 \, c_{Na}{}'' \, E_{Na} = 3,000 \text{ g/cm}^2.$$

This value is characteristic not only for the maximal isometric tension observed in frog muscle but also of that of mammalian skeletal muscle.

Heat and Work

The general relation between free energy, enthalpy and entropy in any physicochemical process is expressed as

$$\Delta G = \Delta H - T \, \Delta S$$

The maximal work obtainable from such a process is given by the decrease of free energy that occurs. Thus

$$\Delta G = -W_{max}$$

In certain processes that produce external work, the enthalpy change ΔH may approach zero. In such an *athermal* process, there may be no change of chemical bond energy. In such a case the work is expressed as:

$$\Delta G_c = -T\,\Delta S_c$$

where ΔG_c is the decrease of configurational free energy, and where ΔS_c is the increase of configurational entropy. For example, a stretched rubber band represents a well-ordered state in which the long chain polymers are in a state of low entropy due to their orientation parallel to the direction of the force of displacement. Release of the force would lead to an increase of entropy and a potential production of external work. The extension or compression of a metallic spring or coil also increases its free energy by an amount of the work done in displacing it from an equilibrium state. This work corresponds to an increase of configurational free energy or to a decrease of the configurational entropy.

The irritability of resting skeletal muscle requires the maintenance of a well-ordered state of low entropy, corresponding to a state of high standard chemical potentials of the inorganic ions. This state is sustained by the intracellular oxidation of substances such as glucose, which is converted anaerobically to lactic acid with an increase of entropy. In the conversion of 1 mole of glucose to 2 mole of lactic acid:

$$C_6H_{12}O_6 = 2\,C_3H_6O_3$$
$$\Delta G = -29.88 \text{ kcal}; \Delta H = -17.67 \text{ kcal}; T\,\Delta S = 12.21 \text{ kcal}.$$

The dielectric energy of the human arm muscles has been estimated in the foregoing as 20 cal (80 cal/kg water). When this is converted rapidly to external work by a sudden contraction in a period of a few milliseconds, it may be assumed that the efficiency of conversion of free energy to work in the athermal period approaches an efficiency of 100%. Then:

$$\Delta G_c = -T\,\Delta S_c = -20 \text{ cal}.$$

At the temperature of 37 °C the production of 20 cal requires the conversion of 0.0016 mole of glucose to lactic acid. Thus:

$$\frac{20}{12{,}210} = 0.0016 \text{ mole}.$$

In the recovery from this process, the myofibrils require a supply of 'negative entropy' obtained from anaerobic glycolysis. Then for the conversion of 0.0016 mole of glucose:

$$0.0016\,C_6H_{12}O_6 \to 0.0032\,C_3H_6O_3: T\,\Delta S_1 = +20 \text{ cal}.$$

In the recovery process (repolarization):

contracted muscle → relaxed muscle: $T \Delta S_2 = -20$ cal.

For the overall process:

$$0.0016 \, C_6H_{12}O_6 + CM \rightarrow 0.0032 \, C_3H_6O_3 + RM$$
$$T \Delta S_1 + T \Delta S_2 = 0$$

where CM and RM denote, respectively, contracted and relaxed muscle. Thus an isotonic contraction which yields 20 cal of external work requires the formation of 0.0032 mole of lactic acid from glucose to yield the 20 cal for the free energy of recovery, ΔG_c or $-T \Delta S_c$. In such a process, the configurational free energy and entropy of the myofibrils remain constant, but there is a loss of 20 cal of free energy and of enthalpy from the nutritive sources. When ΔH and ΔG are exactly equal, the net entropy change of the nutrients, metabolites and respiratory gases is zero. This requires a continuous supply of glucose, fats and other nutrients to maintain an invariant standard state in the cells and tissues. The condition for the maintenance of a steady state in the organism requires:

(1) $\Delta G_{rev} = 0$.

Thus in the standard state, the reversible transport of Na, K, Ca, Mg and Cl is characterized by a value of zero for ΔG. Irreversible transport of O_2, CO_2, nutrients and heat occurs at rates that correspond to the rate of respiratory metabolism. Thus:

(2) $\Delta G_{irrev} < 0$.

(3) Constant dielectric energy, $c_{Na}'' \Delta \mu_{Na}''$, which requires maintenance of a constant state of water and dielectric constant, D'', and the maintenance of the normal state of water and electrolyte balance are implied by conditions 1 and 2.

According to this view of respiration and muscular contraction, the first response of the myofibrils is a rapid phase of isotonic contraction in which the water structure of low dielectric constant is broken. Concurrent events include electrical depolarization, the increase of dielectric constant, and a lowering of the chemical potentials and standard chemical potentials of all ions. This process is attended not only by increases of the solubilities of water-soluble nutrients such as glucose and glycogen, but also by increased rates of glycolysis, and by increases of the respiratory quotient.

Glycolysis depends on the formation of substrates such as glucose-1-phosphate. In this reaction, the high energy bond of ATP is broken to yield ADP. In a related series of reactions:

(a) ATP + glucose = glucose-1-phosphate + ADP

(b) creatine phosphate + ADP = ATP + creatine.

The net reaction (summation of a and b) is equivalent to G + CP = glucose-1-phosphate + C where G, CP and C, respectively, denote glucose, creatine phosphate and creatine. Subsequent steps in glycolysis proceed through a series of phosphate esters including fructose-6-phosphate and terminating with phosphoglyceric acid, pyruvic acid and lactic acid. This series of reactions involves the reversible or irreversible breaking and formation of several kinds of chemical bonds, with production of heat in the various exothermic reactions.

When as a summation of these processes, 2 mole of lactic acid are formed from 1 mole of glucose, ΔG is -29.88 kcal, ΔH is -17.67 kcal, and $T \Delta S$ is 12.21 kcal. In the depolarization phase of contraction, there is a loss of dielectric energy amounting to 20 cal; this corresponds to the value for human biceps and brachialis muscles, which contain a total of about 250 g of water. It corresponds also to the increase of $T \Delta S_c$. Entropy is maintained by the transfer of low entropy from glucose or glycogen to the lengthened or stretched fibers, which are thereby enabled to maintain a constant state of irritability or tonus. Chemical bond energy is made available in a large number of steps in the glycolytic process; this corresponds to the value of -17.67 kcal for ΔH. Entropy is maintained constant by the quantitative transfer of 12.21 kcal of negative entropy in the process of repolarization. The free energy change for the total net process is:

$$\Delta G = -17.67 - 12.21 = -29.88 \text{ kcal.}$$

Thus, per mole of glucose, 17.67 kcal of heat are transported to the environment as 12.21 kcal of free energy (or 'negative entropy') are transported to the repolarized fibers. In the repolarization process, the resting electrical potential is restored, as the value of the dielectric constant of water returns to that of the resting state. The phosphorylation-dephosphorylation processes involving ATP, ADP, and other phosphate esters are involved only in the reversible processes of glycolysis. Restoration of the dielectric energy or configurational free energy of the resting fibers depends on the free energy change, ΔG, of the glycolytic reaction. When this value is taken as -29.88 kcal/mole, the value of $T \Delta S$ is 12.21 kcal. The muscle, including contractile sources of negative entropy (dielectric energy), as well as nutrient sources such as glucose and glycogen, is kept in a constant steady state by the continuous transport of 29.88 kcal of free energy in the form of glucose. This supplies 12.21 kcal of 'negative entropy', measured as $T \Delta S$. Thus for the overall processes of reversible contraction, relaxation and nutrition of muscular tissue:

$$T \Delta S = 0; \text{ and } \Delta H = \Delta G.$$

These are conditions for morphological invariance on a day-to-day basis.

Isometric Heat

In a state of isometric contraction, mammalian muscle fibers develop tension of the order of 3,000 g/cm^2. This corresponds to a decrease of dielectric energy or configurational free energy of about the same magnitude (330 J). Since the conditions of constraint permit no shortening or lengthening of the fibers, the external work is zero. There is, however, an increase of dielectric constant due to the decrease of free energy related to the increase of tension. Thus:

$$dG = -S \, dT - l \, dt$$

At constant temperature:

$$dG = -l \, dt$$

and

$$t = 4.183 \times 8.6 \times 82.65 - 2,970 \, g/cm^2$$

where 82.65 is the dielectric energy in calories, 4.183 is the number of joules in 1 cal, and 8.6 is the product of $10.20 \times \bar{d}w$, where 10.20 converts kilogram-meters to joules, as shown in the previous section. In frog muscle, $\Delta\mu_{Na}''$ is 2.85 kcal; this corresponds to an action potential of about 125 mV (table I).

In an isotonic contraction,

$$\Delta G = -W_{max} = -c_{Na}'' \, \Delta\mu_{Na}^0$$
$$= -82.65 \, kcal/kg \text{ water.}$$

In an isometric contraction that releases the same quantity of free energy, the work of contraction is zero. Then

$$\Delta G = \Delta H - T \, \Delta S_c = -82.65 \, kcal$$
$$= \Delta H - W_{isometric}$$

where the isometric work is zero. Then

$$\Delta G = \Delta H.$$

Since the value of ΔH is -17.67 kcal/mole glucose, the value of 82.65 cal corresponds to the conversion of 4.7×10^{-4} mole of glucose to 9.4×10^{-4} mole of lactic acid per kilogram water.

The maintenance of skeletal muscle in a state of isothermal and isometric contraction thus requires at least three conditions for a constant steady state:

(1) The quantitative supply of glucose at the rate of its combustion in calories per unit time.

(2) The maintenance of a constant intramuscular state of lactic acid by irreversible transport and by simultaneous resynthesis and transport of glucose.

(3) The quantitative removal of the excess heat.

Failure to maintain these conditions would lead to a state of oxygen debt, accumulation of lactic acid, CO_2 and other metabolites, and to an increase of temperature. Maintenance of muscle in a prolonged state of isometric tension and high dielectric constant also tends to produce a state of high utilization of glucose and glycogen, and to a low rate of fat metabolism. This produces a high respiratory quotient. Recovery would require a process of repolarization, lowered dielectric constant and a release of isometric tension. The change of state of water results in its becoming a nonpolar solvent for fats and other lipids, with a lowered value of the respiratory quotient. Complete recovery would require a period of mitochondrial oxidation of lipids.

Chemical Morphology of Neurones

There have been numerous studies of the electrolyte composition and electrical potentials of the giant axon of the squid reported in the literature (*Hodgkin*, 1951). Some of the results are presented in table V. Calculations of the dielectric properties are related to the action potential according to methods described in previous sections. The action potential, E_a, may be related to the change of standard chemical potential of sodium, and this yields the value of the dielectric constant. Values of $\Delta\mu_i^0$ and $(\Delta\mu_i^0)'$ may then be obtained from the

Table V. Dielectric properties and ionization constants of nerve tissues

Tissue	E_a mV	$\Delta\mu_{Na}^0$ kcal	D''	$(\Delta\mu_K^0)'$ kcal	$(\Delta\mu_{Cl}^0)'$ kcal	α_K''	α_{Cl}''
Loligo axon	90	2.08	35	-0.85	0.81	0.018	0.038
Loligo axon	88	2.03	35	-0.52	-0.10	0.019	0.126
Loligo axon	104	2.39	32	-0.38	-0.08	0.026	0.229
Sepia axon	120	2.76	30	-0.42	$-$	0.018	$-$
Sepia axon	124	2.85	29	0.04	$-$	0.030	$-$
Frog nerve	116	2.70	30	-0.16	$-$	0.026	$-$

Action potentials (*Hodgkin*, 1951).

respective values of $\Delta\mu_{Na}{}^0$ and the various values of (r_i) calculated from the analytical data. The method of calculation is shown in the following for *Loligo* axon (*Steinbach and Spiegelman*, 1940, 1943).

(E = 90 mV)	$c_i{}'$	$(c_i)''$	(r_i)	$\Delta\mu_i{}^0$	$(\Delta\mu_i{}^0)'$	$(\Delta G^0{}_{i, Na})'$	$\alpha_i{}''$
Sodium	440	49	0.11	2.08	2.08	–	1.0
Potassium	22	410	19	1.56	−0.85	−2.93	0.018
Chloride	560	40	0.071	1.14	0.81	2.89	0.038

The values of the ion concentrations in blood plasma, $c_i{}'$, are expressed in millimoles per kilogram water. Internal ion concentrations $(c_i)''$ are expressed in the same units. The action potential of the axon, E_a, is 90 mV. This is converted to $\Delta\mu_{Na}{}^0$ by the factor 100 mV = 2.306 kcal, yielding the value 2.08 kcal for the value of the change of standard chemical potential of sodium. From the values of the ionic radii, the values of $\Delta\mu_K{}^0$ (1.56 kcal) and $\Delta\mu_{Cl}{}^0$ (1.14 kcal) are calculated. The value of $(\Delta G^0{}_{K, Na})'$ is estimated from the values of (r_K) and (r_{Na}) obtained from the analytical data. The value of $(\Delta G^0{}_{Na, Cl})'$ is estimated from (r_{Na}) and (r_{Cl}). Then $(\Delta\mu_K{}^0)'$ and $(\Delta\mu_{Cl}{}^0)'$ are obtained from the following relationships:

$$\begin{aligned} (\Delta G^0{}_{K, Na})' &= (\Delta\mu_K{}^0)' - \Delta\mu_{Na}{}^0 \\ -2.93 &= -0.85 - 2.08 \\ (\Delta G^0{}_{Na, Cl})' &= (\Delta\mu_{Cl}{}^0)' = \Delta\mu_{Na}{}^0 \\ 2.89 &= 0.81 + 2.08 \end{aligned}$$

From these results, the values of $\alpha_K{}''$ and $\alpha_{Cl}{}''$ are estimated.

$$(\Delta\mu_i{}^0)' = \Delta\mu_i{}^0 + RT \ln \alpha_i{}''$$

or, at 25 °C

$$\Delta\mu_{Na} = \Delta\mu_{Na}{}^0 + RT \ln (r_{Na}).$$

The calculated values of $\alpha_K{}''$ and $\alpha_{Cl}{}''$ are, respectively, 0.018 and 0.038. The value of the equivalent change of chemical potential, δ, may also be estimated. It is taken as equal to $\Delta\mu_{Na}$, where

$$\Delta\mu_{Na} = 2.08 + 1.37 \log (0.11) = 0.78 \text{ kcal}.$$

The resting potential E is obtained from δ by the conversion factor 1 kcal= 43.4 mV. Then the calculated value of E is −34mV.

Table VI. Resting potentials of nerve (in mV)

Tissue	Resting potential	
	E_{calc}	E_{obs}
Loligo axon	-34	
	-53	
	-56	-61
Sepia axon	-60	
	-81	-62
Frog nerve	-89	-71

Calculated from values of action potentials, E_a, as given in table V. Values of r_{Na} from original data.

Table V presents similar calculations for six different tissues (*Loligo* axon in three different states, *Sepia* axon in two different states, and one preparation of frog nerve). In the amphibian nerve, the value of (r_{Na}) is 0.34, and the action potential is 116 mV (*Hodgkin*, 1951). The calculated value of $\Delta\mu_{Na}^0$ is 2.70 kcal and the calculated value of $\Delta\mu_{Na}$ or δ is

$$\Delta\mu_{Na} = 2.70 + 1.37 \log (0.34) = 2.06 \text{ kcal.}$$

Therefore the calculated resting potential is -89 mV (table VI). Values of the calculated resting potentials for the other tissues are likewise shown in table VI, and compared with observed values.

Dielectric Energy and Physiological Responses

According to the results of tables V and VI, the values of the dielectric constant D'' of certain kinds of axons are of comparable orders of magnitude to those of skeletal muscle fibers of many vertebrates and invertebrates. For example, the calculated value of D'' for frog nerve is 30; this corresponds to a value of 2.70 kcal for $\Delta\mu_{Na}^0$ and to an action potential of 116 mV (table V). For frog skeletal muscle, the respective values of D'', $\Delta\mu_{Na}^0$ and E_a are 29, 2.85 kcal and 125 mV (table IV). Intracellular water of skeletal muscle and nerve in the frog are highly structured, well-ordered and immiscible; both structures deviate widely from the laws of perfect mixing that are characteristic of 'membrane equilibrium'. The action potentials are produced by changes of state of the protoplasm to disordered states of high dielectric constant.

In the behavioral responses of nerve and muscle, dielectric energy of the order of 80 cal/kg water is released, but the physiological functions are entirely different. In skeletal muscle, the energy may be quantitatively converted to the immediate production of external work, amounting to about 350 J/kg water. Under conditions of isometric constraint, however, the response may be only one of increased tension, increased dielectric constant and increased rates of glycolysis and production of CO_2 and heat.

The action potential of a motor neurone, although it corresponds to an almost instantaneous electrical depolarization and very rapid repolarization, produces no mechanical responses of contraction or tension within the nerve fiber itself. The repolarization corresponds to a rapid return of the dielectric energy to the level of the resting state; in a single non-faradic impulse, the dielectric energy corresponding to an action potential of the given magnitude is transferred by an efferent nerve to the motor fiber, where it produces the contractile response.

Motor nerves in the central nervous system seem to be adapted to very fast transmissions of the dielectric energy. This energy produces mechanical effects of external work and tension at points of relatively great distance from the origin of the nerve fibers in the spinal cord or other centers. Actual transfer of metabolic energy from the nervous system to the active muscles is not required to produce the energy of muscular contraction. The work of contraction corresponds to the supply of dielectric energy or configurational free energy that is maintained by muscles in the resting state. This requires a standard state of negative entropy in the muscle fibers, and a constant supply of negative entropy from the intracellular and extracellular nutritive sources.

Resting nerve, on the other hand, requires no comparable supplies of nutritive free energy or 'negative entropy' to maintain normal standard states of irritability or responsiveness. This is also true of the higher centers of the nervous system that are located in the cortical cells of the cerebrum. Mental efforts in the processes of thinking or other kinds of cerebral activity are never characterized by the need for large respiratory supplies of oxygen or by high production of carbon dioxide and heat. Muscular activity requires a high turn-over rate of dielectric energy. In heavy work, this may require an increase of several hundred percent over the basal metabolic work. Although this requires adaptive functions of the central nervous system and rapid processes of recovery of dielectric energy, it does not require metabolic energy of magnitudes comparable to those of muscular work. Intracellular changes of intrafibrillar water of nerve are rapidly transmitted along great distances, requiring only transient changes in chemical morphology of the neurones. These changes are not converted directly to irreversible external or internal work. Consequently they require no large supplies of irreversible nutrient energy, either in brain or in peripheral nerves.

Chapter 7

Central Nervous System

The classification of a living system of cells and tissues as a physicochemical system is necessary to find the laws of its structure and behavior. This may be taken as a fundamental preconception of our study. In the case of the nervous system, this requires classification of the elementary types of cells and their anatomical and morphological interrelations with each other, with muscles or other effectors, and in their relationships with the external or internal environments.

In phylogenetic development, it seems to be established that the most primitive types of biological responses involved only direct environmental stimuli acting on myofibrils or other contractile units (*Parker,* 1918). This is true for invertebrates below the phylogenetic level of the coelenterates, for example. Organisms such as the sponges are capable of producing continuous flow of the environmental sea water through surface pores by way of internal canals leading to external passage through a circular *osculum.* This passage depends on opening or closure of the osculum, which is subject to direct environmental stimulus rather than to central or local nervous control. Control of the passage of the sea water is maintained by regulation of the diameter of the osculum through a muscular sphincter. In colonies of sponges, the flow of current may produce strong effects of turbulence in the sea water at surface levels above the colony. In individual sponges, the movement of water through the pores and communicating channels has been observed directly and shown to depend on the motion of the external sea water, which acts as the stimulus. This kind of direct stimulus of invertebrate muscle below the coelenterate level shows that physiological responses in lower organisms do not necessarily require functions of anything that can be described as nerve cells or neurones.

It is also true that certain kinds of responses in higher vertebrates do not require mediation by nerves. For example, the closure of the iris in mammals is normally brought about by a nervous reflex to the stimulation by intense rays of light impinging on the eye. However, it has been shown that in cases in which there is no possibility of nervous control, the iris closes due to the contraction of its sphincter muscle; this is brought about by direct exposure to intense light.

In the case of stimulation of heart muscle by increases of venous pressure in the circulating blood, it has also been recognized that the origin of the heartbeat may occur directly in certain muscular tissue near the entrance of the right auricle (the sinoauricular node). This leads to a myogenic theory of the contractile process. Mechanical stimulation by changes of venous pressure induces a wave of contraction beginning at that point and proceeding to the left auricle and left ventricle through a process involving periodic contraction and relaxation of all parts of the heart muscle. The presence of certain types of nerve ending in various regions (*Remak*'s ganglion) has tended to support the idea of a neurogenic origin of the heartbeat.

The histological, morphological and neurophysiological evidence does not, however, contradict the evidence for a predominantly myogenic theory of the action of the heart (*Starling*, 1918; *Parker*, 1918). The very complex responses of the heart to changes of pressure and volume of the circulating blood may be regarded as depending not only on direct changes of tension and length of the muscle fibers, but also on nervous stimulation and inhibition. In general, neuromuscular responses in mammals (including man) depend not only on the integrative action of the central and autonomic nervous systems, but also on mechanical changes of tension and length within the muscle fibers themselves (chapter 6). The neuromuscular system of mammals behaves as a unit. Its action depends on physicochemical forces that involve not only striated, cardiac and smooth muscle, but also that depend on the integrative action of the entire system. Tensile forces are exerted not only by muscle fibers, but also by connective tissues; these forces must also affect the overall behavior.

It may be presupposed that the common behavior in all cells and tissues depends on the state of chemical morphology in each type of structure, resting or active. Morphological state implies physicochemical state. In muscle and connective tissue, this presupposes resting and action potentials, as they depend on states of water and electrolytes. This is also true of nerve cells, which have a communicative, integrative function rather than one of independently maintaining mechanical tension or producing external work (chapter 6). All the functions of nerve and muscle depend on changes of state of water, dielectric constant and dielectric energy. Respiratory metabolism at the cellular level also depends on the dielectric properties of intracellular structures.

All elementary properties related to neuromuscular coordination involve morphological properties of the fundamental sets of macromolecular polyelectrolytes, water and electrolytes. The system of cells and tissues in higher organisms respond in various ways to internal stimuli as well as to those that originate in the external physical and biological world. It will be a fundamental preconception of this study that all physiological and behavioral responses imply coordinated physicochemical and morphological responses in the active cells and tissues. Chemical morphology will be understood to include physicochemical state (dielectric properties) as well as chemical composition. Morphological processes universally involve properties such as nervous and muscular irritability, as well as muscular contractility and tension. Within the organism the processes are transmitted and integrated by concomitant changes of water, which exists in labile states of aggregation.

Contraction corresponds to the supply of dielectric energy or configurational free energy that is maintained by muscles in the resting state. This requires a standard state of negative entropy in the muscle fibers, and a constant supply of negative entropy (configurational free energy) from intracellular and extracellular nutritive sources.

Resting nerve, on the other hand, requires no comparable supplies of nutritive free energy or 'negative entropy' to maintain normal physiological standard states of irritability or responsiveness. This is also true of the higher centers of the nervous system that are located in the cortical cells of the cerebrum. Mental efforts in the processes of thinking or other kinds of cerebral activity are not characterized by the need for large respiratory supplies of oxygen or by high production of carbon dioxide and heat. Muscular activity requires a high turnover rate of dielectric energy. In heavy work, this may require an increase of several hundred percent over the basal metabolic rate. Although this requires adaptive functions of the central nervous system and rapid recovery of dielectric energy, it does not require metabolic energy of magnitudes comparable to those of muscular work. Intracellular changes of the intrafibrillar water of nerves are rapidly transmitted along great distances, requiring only transient changes in chemical morphology of the neurones. These changes are not converted directly to irreversible external or internal work. Consequently, they require no large supplies of irreversible nutrient energy, either in brain or in peripheral nerves.

Primitive Innervation

At the lowest level of animal life, at which it is possible to identify innervation of contractile tissues, are found such species as the sea anemones, the jelly fishes and hydroids. Compared with the very sluggish movements of the sponges, the muscles of coelenterates are rather quick; the speed is quickened by the presence of sensory nerve endings at the surface. These may make direct connections with the primitive musculature, and responses to external stimuli then become less sluggish. When compared with the movements of vertebrates and insects, however, the response to stimuli remains slow. In the coelenterates, the effector muscles may be connected directly to the neural receptors. This is the most primitive type of neural organization, involving only sensory nerve endings that receive direct stimulation from the environment.

The anemones are typically marine forms of cylindrical shape. One end, the pedal disk, is attached to a rock or other firm solid embedded in the floor of the sea water. The other end carries a cluster of tentacles that surround the mouth.

This leads downward through a vertical esophagus to a space near the pedal disc, and finally to a digestive cavity. The vertical esophagus is supported by a number of membranes connected to the inner wall of the cylindrical body. The wall of the sea anemone's body is thin and consists of two layers of epithelial cells. These are separated by a third layer of partly secreted substances as extracellular material. The entire outer surface of the cylindrical wall extends upward to the mouth, where it becomes curved inward to connect with the inner surface of the esophagus, which is lined with endodermal tissue.

The very thin ectodermal wall, consisting of three layers, contains various kinds of cells. Among these are a number of sensory cells that terminate at one end in the outer ectodermal layer. These tend to be arranged perpendicularly to the outer ectodermal layer. At the other end they terminate in fine neuro-filaments, which finally form collectively a nerve sheath. Ganglion cells then appear, adding many new fibrils to the aggregate. At a deeper layer than the ganglia are found the effector muscle cells. These are connected to the outer-most nerve endings of the ectoderm. The three distinguishable layers of the primitive neuromuscular system are as follows. The first sublayer includes the terminal cell bodies of the sensory cells of the outer ectodermal layer. The second sublayer includes the nervous sheath formed by the neurofibrils origi-nating in the sensory cells, and also the ganglionic cells, with many branches. The third layer, which may be both ectodermal and endodermal, includes the effector motor cells of the esophagus.

It was found by early investigators (*Hertwig and Hertwig,* 1878) that stimulation of sensory nerve endings in the first sublayer resulted in muscular contraction in the third sublayer. It was inferred that the impulses were trans-mitted through the neurofibrils and ganglionic fibers of the second sublayer. Later investigators modified these conclusions to a certain extent. For example, *Havet* (1901) regarded the sense receptors of the first sublayer to be spread in a diffuse manner throughout the layer, rather than to be concentrated in the oral regions. *Havet* also regarded the ganglia of the second sublayer to consist largely of motor cells, which receive sensory stimuli from the cells of the first layer. These are then transmitted as motor impulses to the effector cells of the third sublayer. According to *Parker* (1918), the cells of the three layers in the outer wall of a sea anemone represent a sequence that is a miniature of the central nervous organs in higher animals. In the latter scheme, a sensory neurone is connected in the spinal cord to a motor nerve, which then leads to a muscle. This is the form of an elementary reflex arc. In certain invertebrates such as the sea anemone, the sensory nerve ending is connected, not directly to the muscle, but through a motor neurone or group of neurones which form a primitive 'nerve center' of the reflex arc.

Although the presence of nervous stimulation and the existence of sensory and motor nerves in actinians such as the sea anemone has been fully established, the histological, anatomical and physiological aspects of the problem may not be fully worked out even at the present time (1978). The problem investigated by the *Hertwigs* should be considered to represent an aspect of the general problem of animal behavior as related to morphological development in very primitive forms of the invertebrate organism.

To quote *Thomas Huxley* (1868): 'It may seem a small thing to admit that the dull vital actions of a fungus or a foraminifer are the properties of their protoplasm and are the direct results of the matter of which they are composed ... (but) the thoughts to which I am now giving utterance and your thoughts regarding them are the expression of molecular changes in the matter of life which is the source of our other vital phenomena.' The 'matter of life' to which *Huxley* here refers is the living substance of protoplasm, which is common to all cells of plants and animals, including the most primitive forms. The living substance, since the 17th century, has generally been recognized to exhibit the properties of irritability and responsiveness to various kinds of stimuli, both in the external environment and internally. An understanding of these properties is essential to any theory of animal behavior at any level of phylogenetic development.

Neurones and Synapses

After the early formulation of the cell theory by *Schleiden* (1842) and *Schwann* (1839), more than a half century was required to arrive at clear and distinct ideas as to the nature of neurones or nerve cells. As early as 1842, *von Helmholtz* demonstrated for invertebrates the relationship between fibrous structures of the central nervous system and corpuscular structures later to be known as ganglion cells. A corresponding relationship between myelinated fibers and ganglia was shown by *Kölliker* in 1844. After these dates it became a problem to establish the integration of the ganglion cells with other parts of the nervous system.

A third constituent of the central nervous system of vertebrates was shown to be the gray matter. In section this material had the appearance of very fine points. About the middle of the 19th century it was known as 'punctate substance'. Later in that period the *Golgi* method of silver impregnation came into use by students of nervous tissue. By 1891, *Kölliker* was able to support the claim that every nerve fiber is at some point connected with a ganglion cell. In the same year *Waldeyer* proposed the theory of the *neurone,* as the first full account of the nerve cell. About 50 years after the cell theory of *Schleiden* (1842) and *Schwann* (1839), it was possible to describe the cellular units of the central nervous system.

After the period of *Kölliker* and of *Waldeyer* about 1891, the ganglion cell of the 1840s or 1850s was recognized as the nucleated body of the neurone, which is the true nerve cell. Two kinds of processes were recognized to emanate from these centers in the ganglion: fine protoplasmic processes and nerve fibers with terminal branched endings. Both types of processes are included in the earlier described material, punctate substance or gray matter. The existence of this material is a possible medium of communication between neighboring neurones.

At about this time it was shown that in the embryological development of nervous tissue, the embryonic *neuroblasts* were in the first stages separated by considerable distances. Later they came together to form the developed neurone. This fusion did not imply the loss of identity of the individual neurones. Rather, the development occurred only to the extent necessary for the transmission of nervous impulses to the branched nerve endings. The system of neurones became finally regarded not as a continuous nerve net, but rather as a system of discrete units maintaining both morphological continuity and discontinuities. Physiological continuity between neighboring neurones was maintained through junctions known as *synapses.* Integrative action of nervous transmission through a system of neurones and their various processes is maintained by conduction through the various synapses.

Anatomic and morphological discontinuity is maintained by the histological differentiation of the various neurones and their fibrillar and protoplasmic processes. Continuity within the nervous system is maintained by the existence of the permanent anatomical and morphological units, especially the *neurones* and their three types of processes. Finely graded and adjusted sensory and motor responses are the function of the elementary reflex arcs, which make their first phylogenetic appearance in certain invertebrates.

The integrative action of the nervous system requires control of the various sensory and effector impulses by conduction or blocking of conduction at the various synapses. According to what has been said in chapter 6, and in previous sections of the present chapter, it may be supposed that physiological responses involving irritability or conduction in any part of the system — neurones or synaptic connections — depend on microscopic or submicroscopic changes of chemical morphology. These changes of state at submicroscopic levels would conceivably involve changes in the dielectric constant of the structured water. Secondary properties would involve standard chemical potentials of the inorganic ions, electrical polarization and depolarization, and changes of dielectric energy. The same principles of irritability, conduction, and electrical polarization in skeletal muscle likewise depend on changes of chemical morphology that involve the dielectric and electrical properties of water and the inorganic ions.

Reflex Action

Although there may be elementary rudiments in invertebrates, the central nervous system makes its first definite appearance in the segmented spinal cord and in the brain of higher animals. In vertebrates such as fish, amphibians and mammals, the fundamental anatomical unit of the central nervous system is the neurone; the functional unit is the reflex arc, which contains effector and motor neurones, connected by various processes in a ganglionic center that includes fibrillar processes in addition to protoplasmic gray matter. Even in the vertebrates, the dorsal root ganglion is not strictly a part of the central nervous system. It occurs rather as a synapse between at least two neurones. In many cases the effector neurone that activates muscle is connected to a number of receptor neurones through their synapses. The effector neurone may thus unify and coordinate the impulses arriving from many receptors. This principle is that of the final common path (*Sherrington,* 1906).

In man, the 'knee jerk' represents a well-defined and easily studied type of spinal reflex. The extensor muscles of the knee contract when the tendon below the patella is tapped in a certain way. By timing the speed of the reflex, *Jolly* (1910) determined the 'synapse time' of the reflex as 0.0021 sec. As opposed to the 'flexor reflex' of the same joint, the knee jerk reflex involves only one synapse. The flexion reflex requires two synapses and three neurones, two of which are receptors. The motor neurone represents the final common path. Thus the synapse time of the flexor reflex is 0.0043 sec compared to 0.0021 sec for the knee jerk. In man and other mammals responses to stimuli are very rapid, of the order of milliseconds. In the knee jerk, the total latency is 0.0055 sec. This is divided into the following responses: afferent endings, 0.0005 sec; conduction time, 0.0014 sec; motor endings, 0.0015 sec. In the flexion reflex, total latent time is 0.0106 sec; response of afferent endings and motor endings is 0.0044 sec, and conduction time is 0.0020 sec. Thus conduction through the afferent and efferent fibers of the reflex arc requires only about 1 or 2 msec through the synapses.

Tendon reflexes, in general, require only very short reaction times. They are therefore among the simplest kinds of spinal reflexes, and may involve only two or three neurones and one or two synapses. In other reflexes, additional neurones and synapses may be required to complete the reflex arc. According to *Sherrington* (1906) the 'scratch reflex' of the dog consists of the following elements:

(1) The receptor neurone from the skin of the back to the gray matter of a spinal segment of the shoulder region.

(2) Connection of the gray matter with a long neurone to the spinal cord, passing to segments of the hind leg.

(3) A synaptic connection to the motor neurone that sends axonic impulses to the flexor muscles, which perform the scratching movements.

In *Sherrington*'s terminology, the motor neurone and its axon form the final common path of the reflex arc. The precentral part of the arc leading from the receptor endings is the *afferent arc*. The entire reflex arc includes sensory endings in the skin or at the roots of the hairs terminating in motor endings in the skeletal muscle. Sensory endings respond to the bites of fleas, for example, of which many dogs are victims. These endings may innervate considerable areas of the sensory regions. The response to impulses in the final common path are not commonly of a highly discriminatory nature, but like other reflex responses, they are rather crude.

The motor responses are usually directed to general regions of the body rather than to very specific points or areas. In general, the responses are crude when compared to the finely discriminated responses that result from the integrative action of the entire nervous system, including the various regions of the brain. These highly integrative responses involve higher centers. In these centers the simple principles that apply to reflex action in the lower spinal centers may not be applicable without extensive modifications.

Cerebral Localization

The systematic study of cerebral localization of motor and sensory areas in the cortex may be considered to have begun with the early studies of *Fritsch and Hitzig* (1870). It was shown that certain movements in various parts of the body were evoked by electrical stimulation of corresponding areas of the cortex. These cortical regions have been called 'motor areas'. However, it should not be inferred that these areas correspond in a one-to-one relationship with definite motor centers such as those located in the spinal motor neurones in the ventral horns of the gray matter. Electrical stimulation of the motor areas of the cortex appears to act on some part of a system of neurones belonging to the afferent side of a lower center leading to the final common path of a particular reflex arc.

Brown and Sherrington (1906) showed that destruction of the motor cortex produces no permanent paralysis of any movements in the parts that correspond to the affected 'area'. Paralysis of the right arm of the chimpanzee was induced by removal of the arm area on the left side of the brain. Recovery of function occurred within less than 5 months. At the end of that time both arms functioned equally well. It was shown that recovery was not caused by regeneration of the area that had been previously removed. Other alternative explanations were investigated and rejected. The only valid explanation seems to be that recovery implied the transfer to a new 'motor area' on the left side of the cortex.

In a later series of studies, *Brown and Sherrington* (1912) applied electrical stimulation to two critical points. One point corresponded to primary flexion of the elbow; the second gave an extensor response. Depending on the experimental conditions, very complex responses were obtained. These included latencies, rebounds, and mutual relations of great complexity of the responses of the arm muscles to cortical stimulation as compared with the highly integrative action of the nervous system in coordinating muscular activity in the normally functioning animal. Results of many investigations since the early studies of cortical stimulation agree that the motor and sensory 'centers' are not located in precise points or even in exactly definable areas. It appears that the areas or regions of effector stimulation can be described as 'labile' or 'mobile', and that the responses depend on many various conditions. These include 'facilitation', or previous treatment, 'reversal of response', and 'deviation of response'. This applies both to precentral and postcentral stimulation, either sensory or motor (*Head,* 1923; *Penfield and Roberts,* 1955; *Penfield and Rasmussen,* 1959).

It is evident from the independent studies of many different investigators of the brain and central nervous system that highly integrated behavior of the mammalian or human organism involves coordinated interactions of all the special senses which relate animal behavior to events and processes in the external world. In higher animals, integrated behavior not only requires responses to these events and processes, but also a high degree of independence from the cosmic environment and its various distractions. This implies a small number of degrees of freedom in the internal physicochemical system of cells and tissues, including the nervous system with its various kinds of neurones, synapses and reflex arcs.

Independent existence of the organism, according to *Bernard* (1878), attains its highest development in those species (mammals), which tend to have complete control of the milieu intérieur. Homeostasis, or physiological invariance, depends on thermodynamic conditions of invariance and constraint that apply to protoplasm and extracellular substances in all their forms. It also depends on physiological responses that are integrated by brain, central and autonomic nervous systems, sense organs, and endocrine secretions. Integration also depends on coordination of circulation, urinary secretion, nutrition and respiratory metabolism (*Joseph,* 1973). Animals are not mere automata whose behavior depends only on mechanical relations with the universe. This is an extreme view, held by materialists of the 17th and 18th centuries, but not lacking advocates in certain quarters up to the present time. The belief derives from certain aspects of 17th century Cartesian dualism, which holds that reality in the physical world is reducible ultimately to simple location of material particles in space and time. Scientific materialism of this crude form is untenable in the 20th century. The climate of opinion is now much more favorable to freedom of the will from external processes in the cosmic environment.

In the animal and vegetable kingdoms only the lowest forms of animals have failed to acquire a certain independence from the cosmic environment. These forms belong either to *vie latente* or to *vie oscillante.* At the other extreme of living forms are the higher vertebrates (mammals and birds), in which behavior is not confined to simple reflex action, tropisms or forced movements, as in the sea anemone or the jelly fish, but in which it is highly integrated by the central nervous system acting in coordination with all other systems of cells and tissues, including the most primitive types.

Table VII. Dielectric properties of human brain

	Dielectric energy cal/kg water	c_{Na}'' mM per kg water	$\Delta\mu_{Na}^0$ kcal/mole	D''
Whole brain	50.05	71.5	0.70	58
White matter	49.60	97.2	0.51	61
Gray matter	55.30	99.5	0.56	59
Loligo axon (ave.)	106	66.7	2.17	34
Sepia axon (ave.)	134	60.0	2.80	30

Dielectric Properties

By means of formulae that have been developed in previous chapters, it is possible to estimate dielectric properties of cells and tissues of which the electrolyte composition is known in the standard state. According to equation 6.10, for example:

$$c_{Na}'' \, \Delta\mu_{Na}^0 = RT \, [\Sigma \, c_i' - \Sigma \, (c_i)'']$$ (7.1)

where c_{Na}'' is the concentration of sodium ions in the cellular or tissue phase, and where $\Sigma \, (c_i)''$ denotes the sum of the concentrations of sodium, potassium and chloride in the same phase. The change of standard chemical potential of sodium is expressed as $\Delta\mu_{Na}^0$, and the product of this and c_{Na}'' yields the dielectric energy in calories per kilogram water. The summation $\Sigma \, c_i'$ denotes the total ion concentration in blood plasma; the normal value may be estimated as 0.315 mole/kg water.

The values of the dielectric properties of human brain, white matter and gray matter, as obtained from data of *Widdowson and Dickerson* (1964), are given in table VII. Values for giant cephalopod axons (loligo and sepia) are included, based on data given by *Hodgkin* (1951). Values of $\Delta\mu_{Na}^0$ are estimated from the action potential, as previously described.

The value of D'' for whole brain is 58; this agrees quite well with values estimated for white matter and gray matter. It likewise agrees well with values previously estimated for human brain in the fetal and newborn states, as well as with those for pig brain in all states of development. Thus it is evident that the dielectric properties of whole brain are fairly independent of age in fetal, post partum and adult brain. Values obtained for rat brain are likewise in good agreement with this result.

When compared with cephalopod axioplasm, however, the values of D'' are high. In mammalian whole brain D'' is usually between 55 and 60, as compared with values of about 30–35 in cephalopod axons. This may indicate a considerable difference with respect to the nature and composition of the protoplasmic and extracellular substances. Brain contains high concentrations of phospholipids such as lecithin and cephalin, as compared with the giant axon, which is probably more representative of the protoplasmic composition of the intracellular phase of neurones. The sodium concentration of human white matter and gray matter is higher than that of the axioplasm. This also could represent a higher ratio of extracellular to intracellular substances than is found in the squid axon. Corresponding to the low values of D'' for the giant axon, the values of $\Delta\mu_{Na}^0$ are high, 2.17 and 2.80 kcal, respectively, in loligo and sepia. Dielectric energies are also high, 106 and 134 cal/kg water. Values for human brain are of the order of 40–50% of these values. Of the various parameters characterizing the two tissues, intracellular sodium concentration, c_{Na}'', appears to be the most constant.

In the squid axon, the values of the dielectric energy are of the orders of magnitude that are found in mammalian skeletal muscle, as well as in the invertebrate type (table VII). Propagation of the motor nerve impulse may be regarded as a progressive flow of dielectric energy from the cell body or ganglion to the synapse. This is attended by the action potential (of the order of 100 mV), which is recorded electrically as a very rapid depolarization and repolarization. The dielectric energy (of the order of 100 cal/kg water) is irreversibly transported to the synapse, resulting in a release of dielectric energy of the same order of magnitude in the muscle fibers.

Action potentials in skeletal muscle are of the same order of magnitude (120 mV) as those in the motor neurone. Irritability and responsiveness in both muscle and nerve depend on maintenance of states of negative entropy (high configurational free energy). This state is maintained in both cases by nutritive supplies of negative entropy, but the requirements for muscle are much greater than for nerve, because of the conversion of the dielectric energy to work by the muscle fibers. It would follow that synaptic conduction would also depend on maintenance of a labile supply of configurational free energy. If so, this implies that neuromuscular behavior depends everywhere on labile states of aggregation of water. Wherever these states become 'frozen', neuromuscular responses are blocked. Changes of state of water, polyelectrolytes and electrolytes should be regarded as the elementary responses to all types of stimuli. This elementary response is common to all animals from the sponge to man. In the lowest forms it may involve only direct stimulation of muscle cells. In higher forms it occurs as very complex central and peripheral responses that are integrated with the behavior of the organism as a whole.

Chapter 8

Sensation and Perception

There is abundant evidence from the literature, art and the archaeological reconstructions of early civilizations that the conscious appeal to the five senses was dominant in all religious and secular aspects of life. The Old Testament is rich in allusions to the religious and secular use of perfumes, incenses and fragrant odors of all kinds. Frankincense and myrrh are frequently mentioned. It also refers to the use of wines and other fermented liquors, as well as to the use of spices and exotic foods. Appeals to the sense of sight were made through the artistic or ritual use of jewels, colored glass and ornaments, and in the arts of dyeing and decoration. In some form or other, the appeal to the sense of touch was made in the texture of clothing and in the tactile surfaces of architectural and sculptural stone or marble. Abundant evidence also is available as to the ritual use of music and the dance in all historical periods. Concurrently there were what we may call puritanical repressions of Orpheus and Terpsichore, as well as of all the other main appeals to the physical senses.

Everywhere and at all times there seem to have been permanent conflicts between Eros and Thanatos. A modern writer, *Norman O. Brown,* has expressed this in the title of a book, *Life Against Death.* Naturally, the fine arts are on the side of Life in this eternal struggle. Empiricism, insofar as it represents only the rationalistic, practical or utilitarian aspects of a civilization, is never on the side of Death, but it may be on the side of a mechanistic, dualistic or Cartesian version of Life. It may not be entirely accidental that the period of *Descartes'* life (1596–1650) coincided with a period of religious and political repression in Europe. It was followed by a nonrepressive period of liberalism and enlightenment, in which the influence of *John Locke* (1632–1704) became predominant.

In the 17th century, the extreme rationalism of *Descartes* appeared not only as a revolt against medieval Aristotelianism, but also against any side of life that was not strictly mathematical. As emphasized by *Gilson* (1938), Cartesian mathematicism marked not only the transition from medieval Aristotelianism to the modern world, but also a revolt against the views of such a thinker as *Montaigne,* who was representative of a type of skepticism characteristic of the late Renaissance. As a transition figure between Medieval or Renaissance thought and our own period of mathematical and symbolic dominance, *Descartes* was perhaps the prime mover in the mode of thinking now known as Cartesian dualism.

Primary and Secondary Properties

When *Descartes* relieved a rather painful period of skepticism by an article of faith 'I think, therefore I exist', he was expressing a logical or metaphysical presuppostion that can be symbolized as

$T \supset E$

where T denotes 'thought', and E denotes 'existence'. The symbol $\supset$ may be read as 'implies'. *Descartes* thus seems to have accepted the preconception that thought implies existence. This would presuppose thought as primary and existence as secondary. In the geometrical sense, his 'clear and distinct' idea of matter required the fundamental notion of extension in space. The basic weakness of the *method* was soon discovered by a number of his contemporaries.

Descartes was unable to bring the realm of sensations into the realm of thought. This would have prevented him, in the absence of a definite sensation of pain, a severe toothhache, for example, from *thinking* the pain and its related discomforts. In a dark room he would not have been able to *think* the color *green* or the color *yellow* or of any colored painting at all. His failure to include the sense organs in his concept of mind led one of his contemperaries, *Leibniz*, to remark: 'At that point, Monsieur Descartes withdrew from the game.' *Descartes* probably relieved his own skepticism by his principle. Subsequent generations of philosophers both in England and on the continent were influenced by the method, but its greatest successes were in pure mathematics. It was not successful in applied mechanics, where the principles of force, motion and acceleration were later established by *Newton.*

Cartesian methods were least successful in the biological sciences, where the great 17th century advances were made by such men as *Harvey* and *Malpighi*, who worked experimentally rather than according to a preconceived 'method' that relied mainly on clear and distinct ideas. These would become accessible only after the laborious efforts of the direct observers, who must always be critical of the data that the mind later accepts or rejects as 'the facts'.

Let us express this attitude in the symbolic form:

$E \supset (S, T)$

Existence (E) of both subject (the observer) and object, as a section of the external world, implies integration of sensory impressions (S) and thought (T) of the observer. Thus S and T are interdependent as members of a set of all the integrated activities of the sense organs, the central nervous system, and the body and mind of the observer. Then

$$(P_1, P_2) \supset (S, T)$$

where P_1 is the set of all relevant processes in the external world, and P_2 is the set of coordinated processes in the body, with its receptor nerve endings and sense organs. This manner of thinking of the relations between S and T is basically nondualistic.

Locke expressed this in his particular form of empiricism, which had great influence on his contemporaries everywhere, and which ultimately became one of the fundamental presuppositions of all branches of science. According to *Locke,* scientific knowledge originates in the sense impressions that reach us from direct observation of the various aspects of the physical world. It is therefore *a posteriori* rather than *a priori* knowledge. The ideas are not necessarily clear and distinct, especially in their early formulations. We observe directly the secondary rather than the primary properties of the external objects; these are the properties that reach the sense organs or the instruments of observation. However refined, these instruments remain fundamentally extensions of the sense organs. The electron microscope or the X-ray diffraction pattern are basically *optical* methods, as are the telescope, the light microscope, or various other systems of mirrors, lenses and apertures that focus light on photographic plates or on the retina.

Then, according to *Locke,* we observe directly only the secondary properties of the external world. The primary properties are inferred according to our intuitions, guesses or hypotheses, which may finally be developed into very elaborate theories regarding the primary entities. These may remain inscrutable, unobservable or inconceivable for very long periods of time. The secondary properties for most of us at most periods of life, constitute the world of colors, sounds, tastes and impressions which is our abode when we are not absorbed into the Cartesian world of pure thought or abstraction.

Like the external world of physical objects, the world of protoplasm, life slimes and living cells and tissues reaches us through sense impressions that originate in the secondary properties of living matter. If we make the absolute presupposition that the primary function of any organism is to exist, this presupposes its genetic origin as a member of a definite phylum, class, genus, and species. In all animals, existence is phylogenetic and ontogenetic. It presupposes a long sequence of coordinated biological processes that finally determine its fundamental chemical morphology as a set of cells and tissues. All other properties such as behavior, contractility, irritability, metabolism, nutritional requirements, adaptability, and so forth are secondary. Without a very long and detailed study, including simple observations as well as the use of the most exact physicochemical methods, the primary properties may remain inscrutable for centuries. Development of the appropriate scientific theory requires not only extensive experimental observations, but also the development of theoretical schemes of energetics and organization. This includes microscopic and sub-microscopic organization of entities that may not be conceivable as physically observable properties. This is especially true of processes of thought or behavior. Our understanding of such processes may necessarily remain largely intuitive or conventional until some remote future time.

Optic Nerve and Visual Organization

As a system of medullated tubes, the optic nerve is a central path organized like the white matter of the spinal cord or cerebrum. Because of the organization and because of the absence of the Schwann cells of peripheral neurones, many of the reactions of the optic nerve and retina resemble those of medullated fibers of the cord and cerebrum rather than those of sensory fibers of other types.

The optic nerve in man consists of about a million nerve fibers organized to transmit nervous impulses from the retina to various regions of the brain, where they are analyzed and resynthesized in various ways to produce perceptual images of external relations of things in space. Corresponding spatial patterns are registered mainly in the superior colliculus of the midbrain, in the lateral geniculate body of the thalamus, and in the occipital lobes of the cerebral cortex. It is the function of these various central regions to analyze and reintegrate the sensory images of the external world as received by the retina. Finally, the central nervous impulses spread through the cortex and are integrated centrally with neurophysiological activities of other cerebral centers by means of complex associational processes. In this way, visual perceptions of spatial objects and colors are brought into relation with all other perceptions of sound, taste, odor and touch, as well as with feelings of pain or pleasure.

As the result of certain kinds of observation, it has been ascertained that perception of the visual field of external objects, colors and light is something that involves the past history and experience of the perceiving individual as well as the anatomical relations of retina, optic nerve fibers and coordinating centers of the brain. Accurate and highly discriminative perceptions require behavior habits that are not immediately learned. Possibly the process is analogous to those previously described as 'facilitation' for various kinds of sensory or motor stimulation of precentral or postcentral areas of the cerebral cortex. The learning process must involve morphological and anatomical as well as neurophysiological development.

The process of learning that is involved in the visual discrimination of external objects has been discussed by *Waddington* (1967). The sources of observations are cited — clinical, psychoanalytic and psychological (*Balint,* 1959), and direct observations (*Young,* 1966). *Balint* refers to a very early period of life in which there is complete harmony between the individual and his surroundings. He has not yet learned to differentiate himself from the external world or to distinguish one external object from another. Exact boundaries are not clearly defined. 'At this stage of development there are as yet no objects, although there is already an individual, who is surrounded, almost floats, in substructures without exact boundaries.' Such individuals live in a harmonious mix-up. Psychoanalytic study shows that detachment of the ego from the external world occurs at a later stage.

Recovery of human beings from certain rare forms of blindness has made possible the direct observation of individuals, earlier deprived of visual perceptions, after they have been exposed to the external world of colors, light and sensible objects. At first, only a spinning mass of lights and colors can be distinguished. External objects can be visualized, identified and named only after a period of a month or more. Experience with individuals who are born blind, but who have recovered their sight, thus shows that visualization is not an innate faculty, but one that is learned by experience.

The evidence cited by *Balint* and by *Young* thus agrees with the 17th century empiricism of *Locke,* who held that our ideas are not innate, but are acquired through experience and training of the human understanding, including the faculties of sensation and perception. At a very high stage of development, these faculties are brought into consciousness and unified with thought and emotions. Only at certain levels of consciousness does this require the association with highly abstract processes of mathematical or logical conceptualization.

The Eye as Camera

The vertebrate eye is made up of a large number of tissues of varied structure and embryological origin. As in the ear, cellular and extracellular structures are coordinated with sense receptors and nerve endings to form an integrated structure. The transparent bodies of the eye include the corneal epithelium and stroma, the aqueous humor, the crystalline lens and the vitreous humor. Cornea, lens and vitreous are solid or semi-solid and heterogeneous; aqueous is fluid and homogeneous. The optical properties of the transparent bodies (indices of refraction, curvatures, focal lengths and positions of optical axes) are almost perfectly adapted to focus images on the retina, where a system of rods and cones conveys stimuli to the optic nerve. Accommodation of the optical system to light intensity and depth of focus is attained by neural and muscular control of the pupil and lens.

Among the structures of the eye, some are of ectodermal origin (corneal epithelium, retina and lens), whereas others are mesodermal (stroma and vitreous). Aqueous humor is a body fluid similar in electrolyte composition to blood plasma, synovial fluid and many others. Like other fluids of the milieu intérieur, it performs a nutritive function. Lens and certain other tissues have little or no capillary circulation. Exchange of nutrients and metabolites is one of the functions of the aqueous humor.

Water and electrolytes are distributed among all the other transparent bodies and contiguous structures. The distribution is not at all uniform. Lens is a rather 'dry' structure which contains only about 65% water. Corneal epithelium, on the other hand, contains about 80% water, and aqueous humor approaches the composition of a blood dialysate. The distribution of sodium, potassium and chloride ions is also non-uniform. The lens is especially rich in potassium, containing a similar concentration of sodium, but only a low concentration of chloride. Corneal epithelium and stroma show electrolyte concentrations rather similar to those of dermal connective tissues, with sodium concentrations of the order of 200 mEq/kg water. No structure of the eye is organized to convert large quantities of energy or entropy in the performance of external or internal work. The energy requirements are small, and tend to be constant. They are adequately met by the nutritive functions of the aqueous humor, which tends to remain in a steady state.

Electrolyte balance in the crystalline lens is described by the following ion concentrations (in mEq/kg water): Na, 108; K, 105; Cl, 17.6. The sum for the three ions is 231 mEq/kg water. When the summation in blood plasma is taken as 315 mEq/kg water, the value of the dielectric energy, RT (0.315−0.231) is found to be 52.08 cal/kg water at 37 °C. This is also the value of $c_{Na} \Delta\mu_{Na}^0$, from which the value of 0.49 kcal is obtained for $\Delta\mu_{Na}^0$. This yields a value of 61 for the dielectric constant, D''. Values of $(\Delta G^0_{K, Na})'$ and $(\Delta G^0_{Na, Cl})'$ are, respectively, −2.06 and 1.31 kcal. As should be expected, the ions are not distributed according to the laws of perfect mixtures or solutions. Since lens is of the nature of a crystalline solid protein, the distribution does not conform to the *Donnan* 'membrane equilibrium'. Since the tissue has no measurable respiratory metabolism, the distribution cannot be explained by 'active transport', selective permeabilities or ion pumps or carriers. It is most readily explained by the physicochemical nature of the lens considered as a semi-solid of dielectric constant of 61. The calculated equivalent weight based on a water content of 608 g/kg tissue is:

$$EW = \frac{392 \times 1,000}{608 \times 0.195} = 3,300 \text{ g/eQ}$$

where the negative colloidal charge is 0.195 eQ/kg water, represented as

$$[(c_{Na})'' + (c_K)'' - (c_{Cl})''].$$

For cornea, the comparable figures are: H_2O, 830 g/kg tissue; $(c_{Na})''$, $(c_K)''$ and $(c_{Cl})''$, respectively, 138.5, 28.4 and 75.9 mEq/kg water. These figures yield the following results as compared with those for lens and adult human dermis (*Joseph,* 1971a, b):

	Cornea	Lens	Dermis
Dielectric energy, cal/kg water	44.64	52.08	13.40
$\Delta\mu_{Na}^{0}$, kcal	0.32	0.49	0.16
D″	66	61	73
Eq. weight, g/Eq	2,250	3,300	7,000

Thus these tissues of ectodermal origin are characterized by relatively unstructured water (dielectric constant 61–73) and relatively low dielectric energy (13.40–52.08 cal). This corresponds to the fact that the structures are largely extracellular and are mainly nonirritable. Their function in the case of cornea and lens is to maintain very stable, nonreactive optical structures of the required indices of refraction and curvature. These properties are summarized as follows:

	Cornea	Lens	
Refractive index	1.37	1.37	
Radius of curvature, mm	7.7	9.2–2.2	(anterior surface)
Radius of curvature, mm	–	5.4–7.1	(posterior surface)
Diameter, mm	12–13	8	

The physical and geometrical properties of the eye are those that focus images of the surrounding environment on the retina, which transfers the picture to the approximately one million neurones of the optic nerve. Adjustments of the eye with respect to light intensity, angular position and distance of visual objects are brought about by sensory and motor reflexes in different regions of the cerebral cortex. Sensory responses follow a regular sequence, ordered in the following way for upper parts of the body – hand, fingers, eye, nose, face, lips, jaw, tongue, throat, etc. (*Penfield and Rasmussen,* 1959). Motor responses have also been studied and follow in part the sequence: hand, fingers, neck, brow, eyelid and eyeball, face, lips, vocalization, etc. Thus the movements of the eye as an optical system are coordinated with many other sensory and motor responses at many centers or regions of the brain. These integrative reactions involving all the sensations and perceptions connect all parts of the body with the external world. The world of sensation and perception brought to *a posteriori* experience through nerve endings is integrated at various levels of the central nervous system. Ulimately the impressions are brought into consciousness and related to sound, speech and various symbolic forms of communication. At these levels of behavior, coordination of mental and perceptual levels of the various human faculties require very difficult processes of learning and habit formation.

The Ear as Musical Instrument

It has been said in an earlier section that music, with the allied arts of color, taste, touch and odor, has been applied in the ritual and secular life of probably all historical periods. These functions could have been only latent in prehistoric life, and might have been brought to consciousness only in the later stages of evolutionary development. Of the various senses, touch seems to have been most primitive because, as we have seen, sensory receptors first appeared in metazoic forms at the level of coelenterates and jelly fishes. Well developed organs of sight and hearing are found only in the higher vertebrates such as mammals and birds. Our domestic animals such as dogs and cats give every evidence of obtaining pleasure and satisfaction from their olfactory organs and the sense of taste. However, it has been observed that in the mating season of all mammals 'the sounds emitted are far from seductive to the human ear' (*Bourlière*, 1953). Thus among the mammals, only man has developed the faculty of speech or verbal communication.

In the early stages of history, this faculty must have been largely utilitarian, but in the periods of high civilization, it certainly gave rise to many pleasurable activities, including poetry, song, music and the various theatrical arts. All of these faculties have to do with the physiological structure of the ear as a receptor for sound stimuli, and with the coordinated neuromuscular control of the organs of vocalization, the mouth, tongue, lips, the vocal cords, and the respiratory passages.

Let us begin by considering musical tones as sound rather than as noise or other audible disturbances which may be perceived as harsh or unpleasant warnings. Tuning forks are used by physicists, musicians and musical craftsmen to establish standard tones of known pitch. A standard pitch corresponds to a certain definite frequency of vibration. When the fork is set into vibration, longitudinal vibrations of the same frequency are transmitted through the air at a speed of about 1,100 ft/sec (12.5 miles/min). The sound waves travel in all directions and at every emitted frequency, thus advancing with a spherical wave front. This sets up vibrations at all hard or elastic surfaces in the path of the spreading wave, causing reflection of the sound in other directions. Thus proper acoustical design of an auditorium requires a judicious distribution of the surfaces and a well planned arrangement of drapes or other nonreflecting surfaces that tend to deaden the tone. Musical instruments, ranging in tone from violins and cellos to trumpets and drums are also resonators.

An important standard pitch that corresponds approximately to middle C on the piano keyboard has a frequency of 256 vibrations/sec. This resonates with its octave (512 vibrations) and with the Cs at higher parts of the keyboard (1,024, 2,048, 4,096). The highest of these notes is near the frequency of the highest note of a modern pianoforte (4,138 vibrations/sec: 4 octaves above middle C, International Pitch). Three octaves below midde C, the frequency of the vibrations is 32/sec, with corresponding frequencies of 64 and 128 in the succeeding octaves. If the note of G (called the 'dominant' of C) is sounded on the keyboard, its frequency corresponds to 48 vibrations/sec. Then the corresponding series of octave of G is represented by the frequencies: 48, 96, 192, 384, 768, 1,536, 3,072.

Each of these is in the ratio of 3 to 2 as compared to C in the same part of the keyboard. When middle C is sounded on the keyboard, the following frequencies are also produced as overtones:

(256)	512	768	1,024	1,536	2,048	2,304	3,072	4,096	4,608
(C)	(C)	(G)	(C)	(G)	(C)	(D)	(G)	(C)	(D)

The tones that correspond to frequencies 2,304 and 4,608 represent D on the scale. 4 octaves below 4,608 corresponds to a frequency of 288. This is the note above middle C. According to a standard known as International Pitch, the A above middle C corresponds to a frequency of 435 vibrations/sec. This leads to the following series of frequencies for the successive notes above C:

258.55	290.33	325.88	345.26	387.54	435.00	488.27
(C)	D	E	F	G	A	B

The ratio 387.54 to 258.55 turns out 1.499 rather than 1.50, which would be the interval of a perfect fifth. This is the ratio of G to C on an 'equally tempered scale'. As is well known, the method of equal temperament was adopted as the standard for the 18th century clavichord. This enables the composer to write in all 24 major and minor keys. It was to prove the practicability of this system that *J.S. Bach* composed the famous 48 preludes and fugues of the *Well Tempered Clavichord* — two books that include each of the 24 possible keys. In the system of equal temperament, none of the 24 major and minor scales is perfect; each is a compromise that gives the composer the advantage of expressing himself in all the possible keys on an instrument with a limited number of notes (88, as on the modern pianoforte).

Each instrument in the orchestra has its own idiosyncracies with respect to limits of pitch, quality of tone, and resonance. Thus the four strings of the violin correspond to G (the lowest one), D, A and E. These are the open tones. D is the fifth above G, A the fifth above D, and E the fifth above A. Each string is tuned to correspond to the first overtone of the next lower string. Thus when the G string is sounded as an open tone, there are resonant vibrations on the other strings corresponding to the various overtones of the fundamental. Many important works for the violin are written in the keys of D or A (the D minor concertos of *Beethoven* and *Brahms;* the A major sonatas of *Brahms* and *Franck,* for example). In these works, the open strings of D and A resonate in the fifths and octaves to produce overtones of the D, A and E strings.

The quality of a violin or other instrument depends on its resonant properties with respect to the vibrating column of air produced by the strings (violin, cello or piano) or by the breath (clarinet, flute or trumpet). The performer on any of these instruments produces a very complex set of longitudinal vibrations that are projected as sound waves directed to the human ear as a sensory receptor. The wave motion of the air consists of a series of very rapid compressions and decompressions of the air. Depending on the complexity of the sound, these compressions and decompressions may occur simultaneously at many different frequencies and intensities.

In a performance of a symphony orchestra, a hundred or more instruments may produce a complex set of vibrations that resonate throughout an entire concert hall. In an ideal situation, the entire hall vibrates in resonance with every instrument. These produce notes ranging in pitch from the lowest on the double bass, contrabassoon or organ pipe to that of the highest string of the harp. The lowest note discernible as a musical tone is said to be C (16 vibrations/sec). The highest note of the modern piano has a frequency of 4,138 vibrations/sec (4 octaves above middle C). The range of the human singing voice is about 60 (low bass) to about 1,300 (high soprano). The upper limit of perception for the human ear is at least two octaves above 4,138 and in many cases is at least 20,000. Certain animals, especially dogs and bats, can detect sounds of very high frequency (ultrasonic waves). Bats make use of this ability to detect high frequency sounds that they emit vocally to be reflected as echoes from surrounding objects. This enables them to fly in the dark in the presence of invisible obstacles.

The sensations of sound, pitch and tone quality in man depend on a series of events that occur in the organ of Corti, an anatomical structure directly connected to the outer ear (*von Helmholtz,* 1862). Longitudinal sound waves enter the ear and set up transverse waves in a structure known as the 'basilar membrane'. This is connected to the vibrating column of air in the cochlea by a system of hair cells that terminate in endings that are found at the outer part of the organ, and that continue toward the base of the cochlea as laterally vibrating hairs connected to the basilar membrane. The hair cells appear to behave as tactile sensory endings that transmit impulses to the basilar membrane. This structure vibrates at all frequencies that occur in the incident column of air that enters the cochlea. It seems to be generally accepted that these resonant vibrations occur as a finely graded series of vibrations that begin at the lowest audible pitches of the musical scale (16 or fewer vibrations and that terminate at high pitches of 20,000 vibrations/sec or higher).

Certain musicians, endowed by nature or by training with an excellent sense of pitch, can discriminate intervals of one-quarter tone or less. They can, for example, distinguish the scale of C sharp from that of D flat. The notes of the piano keyboard are incapable of such discriminations, but violinists or other stringed instrument players can distinguish many more than the 12 tones that can be played within one octave on the piano. Fine discriminations of this kind show that the basilar membrane is sensitive to very fine gradations of pitch.

Although the facts may not be quite so simple, it is generally believed that high frequency sound waves are received at the base of the spiral organ of the cochlea. Low frequencies are registered at the apex, and intermediate pitches between the two limits. The auditory nerve transmits the various impulses to the auditory areas of the cerebral cortex. Tones of high pitch are received as sensory stimuli at the anterior end of the auditory area, with low pitches entering at the posterior part. There appears to be a point for point projection of receptor fibers at localized regions of the auditory region of the cortex.

A projection region for sensory elaboration may be designated by a number known as the *Brodmann* number for the region (*Brodmann,* 1912; *Bailey and von Bonin,* 1951). Auditory perceptual judgment and elaboration has been localized at the *Brodmann* number 41 (*Herrick,* 1953). However, according to *Bailey and von Bonin* (1951), there are serious difficulties in describing the 'cortical architecture' of the various cerebral and thalamic regions. The problem of satisfactory staining techniques for neurones and fibrillar structures in the brain remains in certain aspects unsolved. There are also difficult problems in the preparation and preservation of histological specimens and the control of several difficult procedures. However, the general principle of localization and elaboration of sensory impressions is probably valid, and may be applied to auditory perception as well as to the visual faculties.

Stimulus and Response

There is a general correlation between the narcotic effects of chemical substances and other pharmacological or toxicological effects. In contrast to many toxic properties, narcosis is reversible. It depends on physical as well as chemical interactions beween the narcotic agent and the immiscible phases of the nerve fiber or cell. *Ferguson* (1939) has pointed out that in a homologous series of reagents the effects of narcotics are more closely correlated with the chemical potentials than with the concentrations. When the reagent in a liquid phase is at equilibrium with a cellular phase, the necessary condition requires that $\mu_n' = \mu_n''$, where μ_n refers to the chemical potential of the narcotic; the general condition requires that $\Delta\mu_n = 0$. If the substance is a gas, an additional condition requires that μ_n in the gas phase be equal to μ_n' and to μ_n''. The distribution of the narcotic or anesthetic follows phase rule conditions of equilibrium. In particular the distribution between any two phases conforms, at equilibrium, to *Gibbs'* equation 77, as applied to any non-electrolyte (*Gibbs*, 1875, 1928).

If the anesthetic is a liquid (chloroform, hexane), and if p^0 is the vapor pressure of the pure liquid at the temperature of the experiment, then,

$$\mu_n = \mu_n^0 + RT \ln \frac{p_n}{p^0}$$

where p_n is the equilibrium vapor pressure, and where μ_n^0 is the standard chemical potential in the gas phase. Then for any of the cellular or tissue phases in equilibrium with the gas at the vapor pressure, p_n, μ_n^0 is the standard chemical potential in the physiological phase. For any of the cellular or tissue phases in equilibrium with the gas at the vapor pressure, p_n, $\Delta\mu_n = 0$,

$$\Delta\mu_n^0 + RT \ln S = 0$$

or

$$\Delta\mu_n^0 + RT \ln C = 0$$

where S_n is the solubility in a given phase. This is also expressed by the equilibrium concentration C at the vapor pressure p_n. The value $\Delta\mu_n^0$ represents the difference between μ_n^0 in the biological phase and the value in the gas phase.

A homologous series of hydrocarbons, for example, may be arranged to show a series of isonarcotic concentrations:

Substance	Isonarcotic concentration	$\frac{p_n}{p^0} = kS$
Pentane	0.0052	0.29
Hexane	0.0017	0.34
Heptane	0.00064	0.44
Octane	0.00032	0.82

A parallel column shows the values of p_n/p^0, which are proportional to the solubilities S, in the narcotized cells. As *Ferguson* shows, the values of kS are much more constant than the values of the narcotizing concentrations. Therefore in any series, $(\mu_n - \mu_n^0)$ is of a constant order of magnitude. This shows that the effective concentrations for narcotization or other physiological or pharmacological effects depend on the vapor pressures and standard chemical potentials of the component in all the other phases, including body fluids, cells and tissues.

According to this view, normal neuromuscular behavior requires a normal supply of dielectric energy. This maintains irritability and *tone* in all neurones, synapses and reflex arcs of sensory and motor fibers. This requires maintenance of states of low dielectric constant and high dielectric energy. When this energy is expended in the processes of neural or muscular stimulation, it must be renewed through the processes of glycolytic metabolism. Thus inhibition can occur in either limiting state. The state of low dielectric constant could be sustained by the presence of narcotics such as the aliphatic hydrocarbons. When water is in the state of high dielectric constant, repolarization may be prevented by inhibition of anaerobic glycolysis, with a resulting process of fatigue.

The relationship between the strength or intensity of a stimulus (I) is related to the magnitude of the response (R) by a logarithmic relationship, the *Weber-Fechner* law. Since the 19th century, this has been regarded as a fundamental principle of psychophysiology, although its application to human sensation and perception has often been difficult. If a series of stimuli $(S_1, S_2, ... S_n)$ be arranged exponentially as:

    S: 1 10 100 1,000 ...

then the responses R are proportional to log S, and follow the series:

    R: 0 1 2 3 ...

On the hypothesis that response depends on successive states of polarization and depolarization in the sensory nerve endings, it should be possible to arrive at a rational explanation of the logarithmic relation between R and S. In photochemistry, there is a quantitative relation between the intensity of incident light I^0 and the intensity I after its passage to a distance X from the point of entrance into a given medium. According to the *Lambert-Beer* law:

$$I = I^0 \exp(-a\,x)$$

where a is the absorption coefficient of the medium. According to *Einstein*'s law, the energy of the absorbed light is quantitatively converted to photochemical energy, $h\nu$, where ν is the frequency, and h is Planck's quantum of action. Let us assume that this absorbed energy is converted quantitatively to dielectric energy, $c_{Na}'' \Delta\mu_{Na}^0$ in a process involving a change of state of the structured water in the nerve endings. Then we obtain:

$$c_{Na}'' \Delta\mu_{Na}^0 + RT \ln \frac{I}{I^0} = 0$$

or: transmitted dielectric energy = absorbed photochemical energy.

In any nerve fiber or neurone, the process of transmission of the photochemical or dielectric energy must be followed by a process of recovery, in which:

depolarization energy = repolarization energy.

Thus decreases of dielectric energy in the sensory nerves or in the nervous system as a whole must be accompanied by simultaneous repolarization or recovery processes that depend on metabolic supplies of negative entropy. Dielectric energy and dielectric constant may fluctuate between upper and lower limits in the various structures, but as a whole, the nervous system including sensory endings, reflex arcs, neurones and synapses remains in a steady state characterized by tone and irritability. In the process of visual sensation and perception, for example, the transmissions of photochemical and dielectric energy must be exceedingly small. Nevertheless, perception of the external visual world is very acute and inclusive of very extensive realms of static and dynamic relations.

The cerebral correlation of the visual field, although very extensive, must involve only minute expenditures of dielectric energy. To maintain the steady state of physiological homeostasis, only very small expenditures of metabolic energy (anaerobic and aerobic) are required. Although these expenditures of energy are small, the requirements of the brain and central nervous system for oxygen are very sensitive. A deficiency of oxygen would prevent the recovery of dielectric energy that is normally brought about by respiratory metabolism. Thus inhibitions related to transfer of dielectric energy may occur in two limiting states: inhibitions in the well-ordered 'ice-like' state due to narcosis or anesthesia, and inhibitions of the repolarization or recovery process due to impairment of cell respiration.

When a stimulus is received at a nerve ending, a synapse, ganglion or the body of a neurone, a quantity of dielectric energy is released and transmitted along afferent paths to sensory centers in the cerebrum. Light rays entering the cornea and lens are focussed on the retina, where they stimulate sensory endings of the otpic nerve. Dielectric energy is released at the nerve endings and transmitted along some of the million fibers of the central path. The dielectric energy of whole brain, white matter or gray matter is of the order of 50 cal/kg water (table VII). This corresonds to a value of the dielectric constant of about 60. In the squid axon, the value of D'' is of the order of 30.

The lower value of the dielectric constant indicates a well-ordered state of water, which behaves as a nonpolar solvent in which fats and other lipids are metabolized; in this state glycolytic respiration and respiratory quotient (RQ) are low. Hydrocarbons of the series pentane to octane, when dissolved in cytoplasm, would tend to accumulate in the regions of low dielectric constant. Their narcotic or anesthetic effect may depend on this. In the presence of the hydrocarbon or other anesthetic, water may be prevented from changing to a polar state of high dielectric constant. This would prevent a release of dielectric energy that would otherwise occur when the well-ordered state of 'negative entropy' is stimulated by light. In any state in which the protoplasmic water remains in a state of low polarity, electrical depolarization would be inhibited and the propagation of the nerve impulse would be prevented (*Ecanow et al.,* 1972/73; *Ecanow and Klavans,* 1974). The retinal nerve ending behaves as a reversible 'valve', which may be closed by the action of drugs or anesthetics. Maintenance of homeostasis requires the metabolic transfer of configurational free energy ('negative entropy'), producing the state of low dielectric constant in which water behaves as a structured lipid solvent.

Chapter 9

Biological Energy

Broadly speaking, the energy requirements of human beings, as of all other species of animals, fall into two main categories, with many subdivisions. In the adult state, nutritional energy is required, first of all, to sustain the standard morphological state of all cells and tissues. In the growth period, nutrient energy is required to supply the needs of the cells and tissues for the synthesis of intracellular and extracellular proteins and other macromolecular substances. The other main requirement at any period of life is to meet the needs of the body in the performance of external work, as well as its needs in the various vital processes. These would include the energy required for maintaining the normal states of circulation, respiration, digestion and for the daily activities of the central and autonomic nervous systems.

Maintenance of normal states of homeostasis or thermodynamic invariance on a day-to-day basis implies a continuous distribution of nutrient energy to all the cells and tissues of the body. An invariant steady state of homeostasis throughout the body implies conversion of part of the nutrient energy to configurational free energy, negative entropy or to dielectric energy. The remainder of the nutrient energy is then converted to metabolic heat, which is transported irreversibly to the external environment. Under controlled experimental conditions, this metabolic heat may be transferred to a 'respiratory calorimeter', and quantitatively measured as the standard basal metabolism (*Atwater et al.,* 1897; *Benedict and Cathcart,* 1913).

While nutrient energy, in the form of carbohydrates, proteins and fats, constitutes by far the main supply of external energy, it should not be considered to represent the only source (chapter 8). A certain amount of energy enters the nervous system through the sense organs, particularly through the eyes and ears. Radiant energy enters the optical system through the transparent bodies, including the lens, and is converted to photochemical energy in the retina. A change of state in the sensory nerve endings converts the photochemical energy to dielectric energy, which is then transmitted through the optic nerve and distributed throughout the cerebral cortex, where it makes connections with many sensory and motor centers.

A similar process occurs in the basilar membrane of the organ of Corti when sound waves transmit energy to the sensory endings of the auditory nerve. At any audible frequency, the resonant energy of the vibrations is converted to mechanical energy in the basilar membrane. The mechanical energy is then converted to dielectric energy in the nerve endings and transmitted to sensory and motor endings in the central and thalamic regions. In both auditory and visual processes, optical and auditory impressions are converted to dielectric energy, which is transmitted to central areas in the brain and redistributed throughout the nervous system.

As explained in chapter 8, maintenance of a steady state of irritability or tone in the nervous system requires a continuous supply of nutrient or metabolic energy to reverse the change of state which occurs at the nerve ending or synapse when the protoplasmic water passes from a well-ordered state of low dielectric constant to a relatively disordered state of high dielectric constant.

The well-ordered state is one of high dielectric energy which is released when the nerve ending is stimulated by physical energy that originates in the external world. Maintenance of irritability or tone requires the reversal of the state of high dielectric constant (low dielectric energy). This requires a very small expenditure of metabolic energy derived from nutrient supplies to the nerve endings, neurones and synapses. Every point or region in the nervous system at which the nerve impulse is propagated thus functions as a kind of valve which operates as an oscillator. Conversion of external energy (light or sound) to protoplasmic dielectric energy must lead to a reversible restoration of the protoplasm to the unstimulated state of high dielectric energy. The maintenance of constant tone and irritability requires a continuous supply of nutrient energy that maintains homeostasis throughout the body.

Conditions of Constraint

Basal metabolism in man is of the order of 1,900 kcal/day, or 80 kcal/h (*Morehouse and Miller,* 1967). This represents the enthalpy change of intracellular oxidative processes involving carbohydrates, fats and proteins as nutrients. Invariant metabolic rates presuppose invariant conditions of constraint, as determined by the normal chemical morphology of cells and tissues. As a heterogenous system, the conditions of constraint are determined by *Gibbs'* phase rule, which determines heterogeneous equilibrium in the invariant physico-chemical system (*Gibbs,* 1875; 1928; *Joseph,* 1971a, 1973). The phase rule may be expressed in the form:

f = number of variables − conditions of constraint

where f is the number of degrees of freedom.

With respect to water and electrolyte balance between blood plasma and an intracellular phase, there are five conditions of constraint. These fix the chemical potentials of five components — water and four electrolytes ($NaCl$, KCl, $CaCl_2$ and $MgCl_2$). The intracellular state of each of the five ions must satisfy four conditions:

$$\Delta\mu_{Na} = \Delta\mu_K = 1/2\ \Delta\mu_{Ca} = 1/2\ \Delta\mu_{Mg} = -\Delta\mu_{Cl} = \delta \qquad (9.1)\ (5.5)$$

where δ is the equivalent change of chemical potential of each of the ions (chapter 5). The fifth condition is expressed by the equation

$$\Delta\mu_{H_2O} = 0 \qquad (9.2)$$

The reversible electromotive force E is then determined by

$$FE + \delta = 0 \qquad (9.3)\ (5.12a)$$

where F is the Faraday constant.

These three sets of equations represent *Carnot*'s conditions for an invariant system (*Gibbs*, 1875, 1928). In a system with one degree of freedom, as in the growth, development or aging of mammalian cells and tissues, δ and E may be either univariant or invariant. This is also true of the basal metabolic rate, which would be determined by the state of chemical morphology, as it changes in growth and development. In a univariant system, basal metabolism is a function of morphology and physicochemical state, as determined by the phase rule. After the end of the growth period, E, δ, and metabolic rate would become constant characteristic of the physicochemical state of each kind of cellular structure. The degrees of freedom then approach zero. In that state, δ is a constant, and the number of physicochemical parameters necessary to determine δ is also a constant. In any other state, the number of degrees of freedom is measured as the number of independent variables necessary to determine δ, or the values of the intracellular chemical potentials of the five ions (*Joseph*, 1973).

On a day-to-day basis, the normal adult human body may be considered to be invariant. This requires two conditions: constant basal metabolism and a constant morphological heterogeneous structure. These are the conditions for a *restricted* open system. The physiological system is open with respect to nutrients, oxygen, carbon dioxide, end-products of metabolism and heat. It is closed with respect to fixed anatomical and morphological structures, which must remain in steady invariant states. Thus transport of the inorganic ions and water must be reversible. Distribution of these substances must conform to phase rule conditions of invariance and constraint, which determine resting and action potentials as well as the basal metabolism. Transport of oxygen, carbon dioxide and heat are irreversible processes. As a condition of a steady state of homeostasis, processes of irreversible transport must be correlated with the basal metabolic respiration.

Physicochemical and physiological changes of state must depend on changes of configurational free energy and dielectric energy (chapter 6). Dielectric energy is determined by the relation:

$$c_{Na}'' \ \Delta\mu_{Na}^0 = RT \ [\Sigma \ c_i' - \Sigma \ (c_i)''] \qquad\qquad (9.4) \ (5.1)$$

The intracellular dielectric constant, D'', which depends on the state of intracellular structured water is then given by:

$$D'' = \frac{131}{1.64 + \Delta\mu_{Na}^0} \qquad\qquad (9.5) \ (6.8)$$

where 1.64 kcal is the value of μ_{Na}^0 in the milieu intérieur, in which water exists as a dispersion medium in a system with perfect mixing. From equation 9.4, the value of the dielectric energy of mammalian skeletal muscle is calculated as about 80–85 cal or 335–355 J per kilogram water.

In the arm muscles of an adult man, this amounts to almost 90 J or 9 kg m. This corresponds to the kinetic energy that can be transferred to an external object such as a cricket ball or a baseball (*Hill,* 1944, 1951; *Joseph,* 1971a, 1973). Thus the maximal work of any set of mammalian skeletal muscles is measured by the configurational free energy or the total dielectric energy of the muscle fibers. This can be calculated from the electrolyte distribution, as in equation 9.4 or from the action potential as in equation 5.17.

Distribution of Dielectric Energy

The dielectric energy of human skeletal muscle is about 80 cal/kg water (table IV). Skeletal muscle constitutes about 43% by weight of the adult human body (*Wilmer,* 1940). When the water content is assumed to be 750 g/kg muscle, the total dielectric energy of human skeletal muscle amounts to 2,100 kcal for a man that weighs 70 kg. Brachialis and biceps muscles of one arm have a total water content of about 250 g and a total dielectric energy of about 20 cal. As stated earlier, the dielectric energy of the arm muscles can be converted quantitatively to an external object such as a baseball or cricket ball. In mechanical units, this corresponds to kinetic energy of the order of 85 J.

The dielectric energy of human brain tissue amounts to about 50 cal/kg water (table VII). The average weight of the human brain is about 1,300 g, and the water content is about 80%, or approximately 1 kg. Therefore the total dielectric energy is about 50 cal, or a little more than twice the energy of the brachialis and biceps muscles of one arm. The total number of neurones in the human brain is somewhat less than ten billion (*Herrick,* 1953). Therefore, the dielectric energy of one neurone is of the order of 5×10^{-9} cal. It will be shown in chapter 12 that the dielectric energy of a unicellular organism such as the amoeba is of the order of 7.5×10^{-8} cal. This is somewhat greater than the energy of a single neurone in the human brain. A nervous impulse to mammalian skeletal muscle is of the order of 100 mV, involving a change of standard chemical potential of sodium ion of about 2.5 kcal. This passes through a neuromuscular synapse in the form of a very small quantity of dielectric energy that activates muscular contraction amounting to about 80 cal/kg water. In athletic performance in such an event as the high jump, the flexor muscles of the human thigh rapidly expend about 90 kgm (200 cal) of dielectric energy (*Joseph,* 1973). This exceeds the total dielectric energy of the brain by a factor of four and the dielectric energy of a cerebral neurone by a factor of the order of 4×10^{10}.

The expenditure of very large quantities of energy in the skeletal muscles is thus controlled by very small amounts of dielectric energy originating in the central nervous system. The quality of muscular performance of any kind evidently depends on very accurate coordination and control of the transmission of dielectric energy in the neurones, synapses and reflex arcs of the central nervous system. Recovery of the expended dielectric energy in any part of the neuromuscular system requires efficient supplies of nutrient energy controlled by intracellular respiratory metabolism.

Human Energy Requirements

Energy utilization in various human activities can be measured in several different ways (*Morehouse and Miller,* 1967). One of the most practical methods is the measurement of oxygen consumption in liters per minute. This can be directly converted to kilokalories per minute from the known heats of combustion of the various nutrients. For a man of 70 kg weight, a standard figure of 80 kcal/h is given in the resting state. This corresponds to a basal metabolic rate of 1,920 kcal/day. Other figures are given for work production at various levels of intensity:

	O_2, l/min	kcal/min
Very light work	below 0.5	below 2.5
Light work	0.5–1.0	2.5–5.0
Moderate work	1.0–1.5	5.0–7.5
Heavy work	1.5–2.0	7.5–10.0
Very heavy work	2.0–2.5	10.0–12.5
Maximal work	over 2.5	over 12.5

Moderate work over a period of 10 h thus requires consumption of 600 liters of oxygen. This corresponds to the production of 3,000–4,500 kcal of metabolic energy. This may be compared with the resting rate of 1,920 kcal/day, as given above. The excess over the basal rate thus varies from about 1,080 to 2,580 kcal in a 10-hour period or from 108 to 258 kcal/h. A figure of 10–75 kcal/h can be taken as the magnitude of external work production during a period of moderate work This is based on levels of muscular efficiency ranging from 10 to 30% (*Morehouse and Miller,* 1967).

Basal metabolism in kilocalories per hour or per day is measured as a summation of all the intracellulsr oxidative reactions that involve carbohydrates, fats, proteins and their derivatives. Any such reaction involves not only a change of enthalpy, ΔH, but also changes of free energy, ΔG, and entropy, ΔS. These are related in the following way:

$$\Delta G = \Delta H - T \, \Delta S$$

where T is the absolute temperature. Intracellular oxidations are exothermic, and the values of ΔH are negative. The free energy change of any spontaneous reaction, ΔG, is also negative, even when ΔH is positive. A negative value of ΔG in such an endothermic reaction requires a positive value of $T \, \Delta S$.

For the oxidation of 1 mole of glucose at 25 °C:

$$glucose + 6 \, O_2 = 6 \, CO_2 + 6 \, H_2O.$$

Under physiological conditions, the values of ΔG, ΔH and $T \, \Delta S$ are, respectively, -690.5, -668.5 and 22.0 kcal. For the oxidation of palmitic acid,

$$palmitic \; acid + 23 \, O_2 = 16 \, CO_2 + 16 \, H_2O;$$

$\Delta G = -2,336$ kcal $\Delta H = -2,382$ kcal and $T \, \Delta S = -46$ kcal.

Unlike glucose, palmitic acid is oxidized with a decrease of entropy. As a condition of invariant chemical morphology on a day-to-day basis, it is required that the net entropy change for the two reactions approach zero (*Joseph,* 1971a, 1973). This requires that:

$$a \, T \, \Delta S_a + b \, T \, \Delta S_b = 0$$

where $T \, \Delta S_a$, referring to plamitic acid, is -46 kcal/mole, and where $T \, \Delta S_b$, referring to glucose, is 22 kcal/mole. The number of moles of palmitic acid is denoted as a, and the number of moles of glucose is denoted as b. The condition of perfect energy balance then requires that:

$$\frac{\Delta S_a}{\Delta S_b} = -\frac{b}{a} = \frac{46}{22} = 2.09.$$

In a steady state of homeostasis, a summation of the two oxidative processes yields:

$$2.09 \, C_6H_{12}O_6 + C_{16}H_{23}O_2 + 35.54 \, O_2 = 28.54 \, CO_2 + 28.54 \, H_2O;$$

$\Delta H = -3,779$ kcal, $\Delta G = -3,779$ kcal and $T \, \Delta S = 0$.

Perfect energy balance in a steady state implies that ΔS is zero, as summed over the oxidation reactions. This requires that $\Delta G = \Delta H$, or that the basal metabolic rate be the same as the rate of free energy production. In the state of perfect energy balance, the ratio of b to a is 2.09. The respiratory quotient (RQ) is measured as the ratio of carbon dioxide production to oxygen consumption. According to the foregoing calculations for perfect balance:

$$RQ = \frac{28.54}{35.54} = 0.803.$$

This value is very near the accepted value for the respiratory quotient of normal adult men.

In states of anaerobic respiration, glucose is converted to lactic acid in a nonoxidative reaction:

glucose = 2 lactic acid.

For this reaction, per mole glucose, the values of ΔG, ΔH and $T \, \Delta S$ are, respectively, -29.88, -17.67 and -12.21 kcal, and glycolysis implies a state of negative free energy balance (RQ = 0.803). In a contraction of the human arm muscles that yields 20 cal (83.6 J) of external work:

$$T \, \Delta S_c = 20 \text{ cal}$$

where ΔS_c represents the increase of configurational free energy of the muscles. This corresponds to an increase of dielectric constant and a decrease of dielectric energy. To restore the muscles to the initial state of low entropy, 20 cal of free energy are required from the anaerobic conversion of glucose to lactic acid. The breakdown of 1 mole of glucose yields 12.21 kcal of free energy expressed as $T \, \Delta S$. The remaining 17.67 kcal are transferred to the environment as heat ($-\Delta H$). The transfer of 0.020 kcal of 'negative entropy' to the muscle fibers then requires the anaerobic conversion of 0.00164 mole of glucose to lactic acid. The value is obtained as the ratio of 0.020 to 12.21 kcal.

The heat transferred to the environment is then 0.00164×17.76 kcal, or 29.3 cal. This corresponds to a free energy change of -49.3 cal that occurs during the production of the maximal work of the muscular contraction. The excess of $(-\Delta G)$ over $(-\Delta H)$ indicates that in any period of muscular activity, the body must be in a state of 'negative free energy balance'. This results from an increase of the rate of carbohydrate metabolism referred to fat metabolism in that state. Then the ratio of b to a (glucose to palmitic acid) exceeds the value for the basal state, and the respiratory quotient becomes greater than the basal value of 0.803.

Stimulation of Glycolysis

When a trained athlete at the level of Olympic game competition begins to contract his muscles rapidly at the start of a race or in such an event as the high jump or the broad jump, his actions conform to a prearranged schedule of timing. They also conform to a prearranged set of rules to which he and the other athletes agree. Control of his own behavior begins at the instant he begins his sprint, as in the 100-meter sprint, for example. The timing of this instant is determined by the officials of the game, such as the 'starter', using a loud signal such as a gun.

This is an example of the principle that human muscular behavior is controlled, not only by limited free will of any individual, but principally by conditions that are determined by the external situation. Thus one eats breakfast at 8 a.m., arrives at work at 9 a.m., and begins lunch at noon. Muscular activity is often determined by a prearranged schedule to which the individual conforms. It does not begin in the muscles or in the nervous system, but according to events in the external world, such as the position of the hands on a clock. Even the latter movements are synchronized with events in the solar system, such as the rotation of the earth or the position of the moon.

Accordingly, a large part of human behavior is brought into harmony with earthly or cosmic events. This has been true ever since the evolution of the calendar or other timing devices, such as the sun dial, the hourglass, clockwork, and the curfew. Animal behavior is also, to a large extent, geared to events in the universe such as the sunrise, the sunset, and the seasons of the year. Much of this behavior becomes instinctual, as it originates in geophysical conditions, rather than in the internal neuromuscular processes. Thus it would be fallacious to consider animal or human behavior to be initiated by processes that originate in intracellular respiratory metabolism. These processes, as in the behavior of the Olympic athlete, conform to events in the external world. External events are independent of the conditions of intracellular or intramuscular phosphorylation processes. This is said in order to emphasize that all human behavior from the inauguration of a President to the coronation of a King depends on events in the external world. Many such events are beyond the control of any one human being.

When an athlete, in response to the starter's gun, begins his sprint, there is an immediate increase in his metabolic rate, which may then exceed a value of

12.5 kcal/min, as compared with a basal value of 1–2 kcal/min. In the active muscles, external work is obtained from the configurational free energy of the muscles. This is always maintained in the muscle fibers by cellular respiration, which continuously supplies 'negative entropy' to the muscles, along with irreversible transport of heat and carbon dioxide, and irreversible consumption of oxygen. During the period of muscular activity, in which metabolism may exceed the level of 12.5 kcal/min the muscle is in a state of 'negative free energy balance'. This implies a high utilization of carbohydrates and glycogen as sources of free energy and negative entropy. It also implies a change of state of intrafibrillar water, an increase of dielectric constant, a decrease of dielectric energy, and an increased respiratory quotient.

Let it be assumed that a man performing external work during a period of 10 h expends 2,500 kcal of configurational free energy in excess of his basal metabolism (*Joseph*, 1973). If it is assumed that during this 10-hour period the RQ increases to 0.82, as compared to the basal level of 0.803, this corresponds to an increased value of b/a. Here b denotes the rate of oxidation of glucose, and a represents the rate of oxidation of palmitic acid, expressed as number of moles. At the RQ of 0.82, the value of b/a becomes 2.65 rather than 2.09, as in the resting state. Under these conditions, the value of $T \Delta S$ is 10 kcal, or 1 kcal/h. This is equal to the difference $(\Delta H - \Delta G)$, which measures the 'negative free energy balance' during the 10-hour period of muscular activity.

A period of 'positive free energy balance' is required, in which $T \Delta S$ is -10 kcal. This meets the condition for day-to-day invariance which implies a value of zero for $T \Delta S$. If the rest period is measured as 14 h required for complete recovery of the dielectric energy, it is estimated that the RQ decreases to a level of 0.73. During the 14-hour recovery period, the mean value of b/a is calculated as 0.51. Thus during the recovery period, the muscles shift to a state in which fat metabolism is high, and carbohydrate metabolism is low. The condition for an invariant steady state requires a value of 2.09 for b/a over the 24-hour period. It also requires a value of 0.803 for the respiratory quotient.

During a period of high muscular activity, the value of the intracellular dielectric constant, D'', is increased above the resting level of about 30. The standard state is restored after the recovery period, when the RQ returns to the value of 0.803, corresponding to the value of 2.09 for b/a. Thus the values of RQ and b/a fluctuate as the secular variations of the state of water depend on rates of energy metabolism. An increased value of b/a is produced by an increase in the rate of glycolysis brought about in the following way. Let k_r measure the rate of glycolysis in the standard resting state, and let $\bar{k}_r$ represent the rate in an activated state in which water is in the disordered state of high dielectric constant D''. This is brought about by the process of contraction, which lowers the configurational free energy or dielectric energy of the fibers, as measured by $c_{Na}'' \Delta\mu_{Na}^{0}$.

Water in the state of high dielectric constant would theroretically behave as a solvent for such substances as glucose, glycogen and numerous phosphate esters such as ATP, ADP and creatine phosphate (CP). The initial step in the formation of lactic acid from glucose is the formation of glucose-1-phosphate (GP). This is brought about by the reaction:

glucose + ATP = GP + ADP.

ATP is regenerated in the reaction:

ADP + CP = ATP + C

where CP denotes creatine phosphate (phosphagen) and C stands for free creatine. This reaction would determine the rate of formation of glucose-1-phosphate. Combining the above reactions:

glucose + CP = GP + C.

The rate of this reaction would be proportional to the rate of formation of lactic acid, and to the rate of transfer of configurational free energy and negative entropy of the muscle fibers. In a steady state, it would be proportional to the rate of production of external muscular work. The rate of glycolysis, $\bar{k}_r$, is theoretically proportional to the equilibrium constant K^*, where

$$K^* = \frac{(GPC)^*}{(GP)\,(C)}$$

$(GPC)^*$ is the concentration of an activated complex formed from glucose-1-phosphate and creatine. Thus

$$\bar{k}_r = \frac{k\,T}{h} K^*$$

where k denotes *Boltzmann*'s constant, and h denotes Planck's quantum of action. Then a general expression for $\bar{k}_r$ is of the form:

$$\bar{k}_r = \frac{k\,T}{h} \exp\left(-\frac{\Delta H^*}{RT}\right) \exp\left(\frac{\Delta S^*}{T}\right)$$

where ΔH^* is the heat of activation and ΔS^* is the entropy of activation (chapter 3).

The glycolytic rate constant $\bar{k}_r$ is related to the resting value k_r by the relation:

$$\frac{\bar{k}_r}{k_r} = \frac{(GPC)^*}{(GPC)}$$

where the numerator and denominator refer, respectively, to the concentrations of the activated complex in the active and resting states. An increase of dielectric constant, as in muscular activity, favors solubilization of reactants such as glycogen, glucose, creatine phosphate and ATP. Conversely, a lowering of dielectric constant implies an increase of mitochondrial oxidation of fats, fatty acids and other lipids corresponding to the behavior of water as a lipid solvent. The rates of glycolysis and mitochondrial oxidation thus tend to fluctuate in a one-to-one relationship with the intracellular states of water. Homeostasis on a day-to-day basis requires a null value of $T \Delta S$, which implies constant values of b/a and RQ. These values are, respectively, 2.09 and 0.803. They are constant on a day-to-day basis but fluctuate as the energy requirements of the body change with exercise or other activities.

Efficiency of Muscular Activity

Efficiency is measured as the ratio of energy conversion to energy production, as measured in percentage. For human subjects it may vary from about 5 to 30 %, depending on various factors, such as the work load or the rate of energy conversion. In general, efficiency tends to increase as the rate of work per unit time is increased. Because of the complexity of many of the factors, it is not usually possible to arrive at exact values for efficiency in human subjects. In the following, calculation of the energy production in an athletic event will be attempted.

An athlete running at a speed of 18.9 mi/h consumes energy at the rate of 9,480 kcal/h (*Morehouse and Miller*, 1967). These figures correspond approximately to a speed of 8.5 m/sec, and to an energy consumption of 2.63 kcal/sec. The flexor muscles of the human thigh have a water content of about 2.5 kg. Estimating the dielectric energy as 85 cal/kg water, the value for the flexor muscles is about 2.12 cal or 885 J for a water content of 2.5 kg.

In the 100-meter sprint, an athlete weighing 70 kg attains a speed of 10 m/sec. This corresponds to a kinetic energy of about 3,500 J, or the dielectric energy converted to kinetic energy in four strides, as estimated from the ratio 3,500 to 885 (*Joseph*, 1973). The energy consumption is 2.63 kcal/sec or 26.3 kcal for 10 sec (at the level for Olympic competition). Therefore the estimated efficiency is obtained from the ratio of 850 to 2,630 cal (850 = 212 $\times$ 4). This amounts to an efficiency of about 30% a reasonable figure for a high level of work production.

The figure of 850 cal in a period of 10 sec should be compared with the dielectric energy of the thigh (212 cal) and with the dielectric energy of the human brain (50 cal) (table VII). The total conversion of dielectric energy to kinetic energy is about 850 cal. This exceeds the total dielectric energy of the brain by a factor of 17. Therefore the energy expended in muscular work must be derived almost entirely from nutritive sources in the muscle. Expenditure of nutrient energy in the brain and spinal centers must be extremely small. As in any well-designed machine which combines expenditure of fuel, controlled by central mechanisms, the bulk of the fuel is expended in the external work rather than in the controls. In an automobile, the energy is spent in moving the drive wheels rather than the steering wheel. In a well-trained athlete, the metabolic energy is spent in the skeletal muscles rather than in the central nervous system. Therefore the body behaves as a well-designed intelligent machine, acting with well-defined purposes. It is not an automaton.

Reactivity of Muscle Proteins

The energetics of muscular contraction are largely dependent on changes of distribution of water and on the dielectric properties of intrafibrillar, interfibrillar, and membraneous water in the myofibrils (*Joseph, 1973; Catchpole and Joseph, 1974*). This depends largely on the reactivity of the normal muscular proteins, actin and myosin (*Frey-Wyssling, 1953*). The contractile substance of muscle is contained in the insoluble fraction which remains after dissolution of a soluble protein, myogen, not included in the contractile proteins. Myosin normally occurs in fractions that contain another protein, actin, which exists in both globular and fibrillar forms (G and F actin). Neither myosin or F actin is contractile, but if the two proteins are brought together, they react to form the contractile substance, F actomyosin. There appears to be a stoichiometric ratio of about 2.5 parts myosin to 1 part of F actin in the reactive muscle protein (*Snellman and Erdos, 1948*). Under certain conditions, myosin and actin react vigorously in a contractile process which involves soluble products and electrolytes. The interfibrillar water has an apparent dielectric constant of 80, as compared with a value of D'' of 30 for the normally uncontracted protein (*Joseph, 1973*). The change of standard chemical potential of sodium, $\Delta\mu_{Na}{}^{0}$, is then given by the relation:

$$\Delta\mu_{Na}{}^{0} = 164 \left(\frac{1}{30} - \frac{1}{80} \right) = 2.73 \text{ kcal/mole.}$$

The work content of muscle is then given by the dielectric energy:

$$c_{Na}{}'' \, \Delta\mu_{Na}{}^{0} = 0.03 \times 2.73 = 81.3 \text{ cal/kg water.}$$

Since the normal water content of adult skeletal muscle is about 250 g/kg muscle, the value of the dielectric energy (or hydration energy) is of the order of 20 cal for the biceps and brachialis muscles of an adult man. This is a measure of the work capacity of muscle, as measured by muscular proficiency in a great variety of athletic events, such as the sprint, throwing baseballs or cricket balls, the high jump, and so forth. Dielectric energy is also a measure of isometric tension, and of muscular performance. Thus the normal value of the dielectric energy of mammalian or frog muscle corresponds to a normal isometric tension of about $3,000 \, g/cm^2$ radius (*Joseph,* 1973). The value depends on the labile hydration energy of sodium ions, as it depends on changes of state of intra-muscular water resulting from the reactivity of myosin and actin.

Regarding the functions of the muscle proteins in contraction, the following description can be given (*Catchpole and Joseph,* 1974). 'Regarding mechanical models of muscle that are based on morphology which are currently in vogue, it always has been recognized that muscle has a characteristic form which alters on contraction in an invariant way. In the 17th century, *Leeuwenhoeck* already had discovered cross-striations which have been revealed more clearly by stains and modern microscopes. The electron microscope has revealed a wealth of further detail and finer ordering, while X-ray analysis adds its quota of structure. These macro- and ultramicroscopic structures constitute the "moving parts" of the machine. But these are not the 'origin' of the motion any more than the busy pistons, rings, wrist pins, connecting rods, and crankshaft of a gasoline or diesel engine. In muscle, a transient change in the dielectric constant of water initiates disassembly and energy liberation, followed by self-assembly and energy absorption, in which every order of structure participates.'

Auditory Perception

The following considerations, due to the paucity of exact quantitative figures, are largely hypothetical and speculative, although they are based on observations that were well known to the ancients. Such speculations have been entertained by more modern writers such as *Jonathan Swift,* and by entomologists of an imaginative bent. They are based on the observation of the apparently great muscular strength of such insects as ants and termites, who can be seen to lift and transport objects of great weight as compared to that of the bearer. These impressive feats of strength, on a weight to weight basis, might well exceed the feats of human weight lifters, of draft horses, or even of the elephant.

A rational explanation of the relation of physical work to anatomical size is based on the physical principle that the energy requirements of animals are proportional to the surface area rather than to weight or volume. Surface area, S, is proportional to the square of a linear dimension. Weight is proportional to the volume, V. The ratio of total surface to volume is proportional to the $^2/_3$ power of the mass. This principle may be illustrated by comparing mass and the ratio of surface to mass in a series of organisms beginning with man at a weight of 70 kg. Let us extend the series through organisms A, B, C, in each of which the mass is decreased by a factor of 1,000. A corresponding series of basal metabolic rates is then arranged on the principle that the rate is proportional to the ratio of surface area to volume. Then the following series of values is obtained:

	Mass	S/V (relative)	Basal metabolism, kcal
Man	70 kg	1	2,000
A	70 g	10	200
B	70 mg	100	20
C	0.070 mg	1,000	2

The hypothetical basal metabolism is then proportional to the $2/3$ power of the mass. This decreases by a factor of 10 for each 1,000-fold decrease of the mass. Therefore the basal metabolic rate of a small insect would be disproportionately high when referred to the mass.

The value in calories per gram increases by a factor of about 1,000 as we pass from a large mammal to a small insect. This accounts for the apparent great strength of ants and termites. It would also account for their relatively high speeds of walking or running and for the very rapid movements of bees, ants and houseflies.

In flight, bees emit audible sounds of high frequency, which at close range may be very annoying to man and to his domestic animals. In very small insects such as mosquitoes, the audible movement of the wings can become intolerable. The relatively high pitch and intensity of the sound must correspond to a very rapid movement of the wings. In the case of the bee, the pitch seems to correspond to a frequency of the order of 200 or 300 vibrations/sec.

There are two ways of looking at the origin of the muscular motion. One theory, to which much attention has been directed in recent years, is based on the supposed breakdown of adenosine triphosphate (ATP) or of other phosphate bonds that yield chemical bond energy to the muscle fibers. As presently formulated, this theory requires the operation of 'calcium pumps', 'sodium pumps', and enzymes such as 'ATP-ase' and sodium activated ATP-ase. These sources of bond energy supposedly operate through reversible cycles that require the oxidative rephosphorlylation of ADP to ATP. To explain the origin of the high frequency of the insect wing, it would be necessary to imagine a cyclic source of bond energy operating at a frequency of the order of 200 or more per second. The total bond energy of intracellular or intramuscular ATP amounts only to about 10 cal in the human arm muscles and about 100 cal in the flexor muscles of the thigh. It has previously been shown that in the 100-meter sprint an athlete expends about 26,500 cal. The bond energy of ATP could account for only about 2,000 cal, or about 8% of the total.

An alternative theory would place the origin of the muscular contraction in a neurogenic transport of dielectric energy passing through neurones and synapses of central or local reflex arcs. This is based on calculations of dielectric energy related to the standard chemical morphology of skeletal muscle. This amounts to about 80 cal (335 J) per kilogram water (*Joseph*, 1971a, b, 1973). In previous chapters the calculations have been extended to include nerve axons and human brain, including white and gray matter (table VII). The theory has also been extended to include the transmission of auditory and visual energy impulses through sensory nerve endings. This implies that nerve endings and synapses behave as reversible valves that depend on very rapid oscillations of the dielectric properties of protoplasm (see also earlier sections of this chapter).

It is now evident that we have a theoretical principle that can account for the perception of high frequency sound waves as well as for their emission, as in the flight of insects or in the vocal cords of mammals. The human ear is capable of perceiving frequencies as low as 16/sec and as high as 20,000/sec. This range brings the auditory faculties into perception of frequencies emitted by the vocalizing activities of many kinds of mammals. It also permits the perception of such natural sounds as the chirping of the cricket or the flight of the humming bird. These sounds are produced by changes of the dielectric energy in the muscles of the given species. Thus through their sense organs, all animals enter into mutual ecological relations that involve changes of dielectric energy at synapses in the sensory nerve endings and at the motor endings of the sending apparatus. Auditory communications within any species or between two different species depend ultimately on the transmission of high frequency sound waves produced by oscillations of dielectric energy at the source of the waves, and by corresponding oscillations in the receptor nerve endings and synapses. What we actually perceive when we hear the flight of the bumble bee is the oscillation of dielectric energy of the neuromuscular apparatus. According to this theory, all animal communication depends on reciprocal changes in the state of protoplasm at both ends of the communication system. Communication depends on the exchange of auditory and visual signals.

Principle of Similitude

Problems of force, velocity and motion in many species of mammals, birds and insects are related by a general 'principle of similitude'. The extensive literature on the subject is cited by Sir *d'Arcy W. Thompson* in *Growth and Form* (1944). The general principles, as applied to problems in physics and engineering, may be derived from elementary mathematical relations among such properties as length, area, volume and mass. Biological applications have been appreciated by many physiologists and physicists dating from the time of *Galileo.* They include considerations of the nutritional requirements of mammals and insects, and the relationship between metabolism and speed of movements.

Some of the quantitative applications of the general principles are shown in the following results, adapted from *Thompson.* These describe the relation between body weight and metabolic rate for a number of mammals.

	Weight, kg	Metabolism kcal/kg
Guinea pig	0.7	223
Rabbit	2	58
Man	70	33
Horse	600	22
Elephant	4,000	13

As applied to the flight of birds and insects, the following results have been obtained:

	Weight g	Beats/sec	Force of wingbeat, g	Specific force (F/W)
Stork	3,500	2	1,480	2:5
Gull	1,000	3	640	2:3
Pigeon	350	6	160	1:2
Sparrow	30	13	13	2:5
Bee	0.07	200	0.2	3.5:1
Fly	0.01	100	0.04	4:1

On a weight basis, insects such as bees and flies, according to the principle of similitude have very high metabolic rates, as compared to those of birds or mammals. They also have very high frequencies of wing beats and high specific force of the beat. In accordance with their high metabolic rates per unit weight, their speed of flight and kinetic energy are also very high. These parameters correspond to high rates of conversion of dielectric energy, as determined by the frequency of oscillation. The dielectric energy per kilogram of intrafibrillar water may be of the same order of magnitude (50–100 cal) for many kinds of muscle fibers.

The frequencies of oscillation of insect wings, expressed above as number of vibrations per second, are of the order of 100 (housefly) and 200 (bee). These correspond to frequencies in the audio range. They may be taken as typical for common insects, but higher values may be found in other species. The reciprocal values in seconds per vibration may be called the 'apparent relaxation times' of the oscillation. The values would be 0.01 sec (housefly) and 0.005 sec (bee). Corresponding values for birds would be of the order of 0.5 sec (stork), 0.3 sec (gull) and 0.08 sec (sparrow).

The normal human heart beating at 70 strokes/min would have a relaxation time of the order of 0.9 sec. A man running the 100-meter sprint in 10 sec would contract his leg muscles at the rate of about 5/sec, with a relaxation time of the order of 0.25 sec. Since the human ear perceives audio frequencies of about 20–20,000/sec. the 'apparent relaxation time' of the nerve endings of the ear could be estimated as from 0.05 to 5×10^{-5} sec. Thus, with few exceptions, the frequencies of vibration of the muscles of mammals and birds are not within the audible range. Unlike mammals, insects are capable of producing audible sounds by their organs of locomotion. This is explained by the principle of similitude, according to which their rates of metabolism are disproportionately high.

The relaxation time of protoplasm in the auditory nerve endings is of the order of 0.05 to 5×10^{-5} sec. The latter value, approaching a possible upper limit for the speed of protoplasmic oscillations is of a much slower order of magnitude than those found for aqueous solutions of various kinds of proteins, amino acids or peptides (*Oncley,* in *Cohn and Edsall,* 1943). A few values for these substances will show the typical magnitudes of relaxation times in aqueous solution:

Substance	Relaxation time, sec
Glycine	3.7×10^{-11}
γ-Aminobutyric acid	6.2×10^{-11}
Pentaglycine	36.3×10^{-11}
Myoglobin	2.9×10^{-6}
Pig carboxyhemoglobin	1.3×10^{-7}

All these values are those of amino acids, peptides or proteins in the state of homogeneous aqueous solutions. The fastest of the relaxation times (for the amino acids) are of the order of 10^{-11} sec. The slowest, 2.9×10^{-6} sec (for myoglobin) is slower by a factor of almost 10 than that for the most rapid oscillations perceptible in the nerve endings of the human ear (5×10^{-5} sec). Thus, physicochemical behavior in the mammalian nervous system should be regarded as dependent not only on the size and shape of the macromolecular substances of protoplasm, but also on the state of aggregation of the heterogeneous system.

Energy of Perception

The number of neurones in the human brain is of the order of ten billion, and the total dielectric energy of the brain is about 50 cal (table VII). Therefore the oscillatory changes of dielectric energy at the sensory endings of the auditory nerve must be extremely small. Nevertheless, they are sufficiently sensitive to respond to the vibrations emitted by the wings of a small insect.

An insect of 70 mg weight, as has been shown, may be assumed to have a basal metabolism of the order of 20 kcal/day, or about 1 kcal/h. If some of this energy were spent in a flight of 10 sec, the total energy would be of the order of 0.28 cal/sec. Let us assume that the energy spent in the wing movements is about one-third of the total, or about 0.10 cal/sec. If the emitted sound waves had a frequency of 280/sec, the energy per vibration would be of the order of 3×10^{-4} cal. (for the motion of two wings), distributed over a spherical wave front. The area of the wave front depends on the radius of the sphere, or the distance of the ear from the vibrating wings. Let us assume that 1% of the emitted energy enters the ear to set up vibrations in the cochlea of the receptor. Then only about 10^{-6} cal are sufficient to cause vibrations in the basilar membrane at the frequency of 280/sec. This signifies that the dielectric energy transported through the auditory nerve is 10^{-6} cal/sec.

Since a given sensory nerve ending is required to oscillate in resonance with vibrations of 280/sec, the energy per vibration is 3.6×10^{-9} cal. This corresponds to the conversion of 10^{-6} cal/sec. If the dielectric energy in the nerve ending is 50 cal/kg water, the water content of a given ending would be 7.2×10^{-8} g. Then 1 mole of protoplasmic water in the gray matter corresponds to the quantity contained in about 1.4×10^7 nerve endings. The oscillatory energy entering the system is thus distributed among about 10,000 nerve endings per milligram of protoplasmic water.

Thus it can be said that the motion of a small insect is sufficient to activate nerve endings that contain 7.2×10^{-8} g of water. The amount of sound energy is perceptible even when less than 1% of it reaches the organ of Corti. The energy is distributed over a spherical wave front of considerable area. A large fraction does not enter the ear. In the motion of the wings, only a small fraction of the total energy would be emitted as sound. The total rate of energy production might be of the order of 0.2 cal/sec. If dielectric energy of this order of magnitude is produced in the muscle fibers of the wing, only a small fraction of the total energy would be emitted as sound and a much smaller fraction would reach the ear.

From the rate of energy production, 0.2 cal/sec, it might be possible to estimate the mass of the intramuscular water. Beating at a rate of 280 vibrations/sec, the number of calories per vibration is 7.2×10^{-4}. This represents the dielectric energy content of about 7 mg of muscular water.

Protoplasmic water of the wing muscles changing state at a frequency of 280/sec is sufficient to generate about 0.2 cal/sec. This emits sound energy at the rate of about 0.07 cal/sec at the given frequency. It is also sufficient to activate oscillatory dielectric energy in the auditory nerve endings, amounting to about 10^{-6} cal/sec.

The wing movements of the insect emit much more energy than is required to activate resonating oscillatory frequencies in the dielectric energy of the nerve endings. This amounts to 3.6×10^{-9} cal per oscillation, representing dielectric energy of 7.2×10^{-8} g of protoplasmic water. In the transmitting apparatus, a much larger quantity of intramuscular water is required to generate energy in the basilar membrane of the organ of Corti at the rate of 10^{-6} cal/sec. This would be sufficient for the dielectric energy of 10^{-4} nerve endings at the rate of 10^{-10} cal per ending per second. The mass of protoplasmic water in the wing muscles exceeds that of the sensory nerve endings by a ratio of 7×10^{-3} to 7.2×10^{-8}. This is a ratio of about 100,000. The dielectric energy of the wing muscles exceeds that of the sensory endings by a factor of about 10^5. Nature compensates for this difference by a great proliferation of sensory endings in the receptor apparatus.

A spherical mass of water weighing 1 g has a volume of 1 cm^3 and a surface area of 4.8 cm^2. The ratio of surface area to volume is 4.8 cm^{-1}. The ratio S/V increases by a factor of 10 for each 1,000-fold decrease of volume. If there are 1.4×10^7 nerve endings, each of which has a content of 7.2×10^{-8} cal of protoplasmic water, the total surface area of the water would be about 1,000 times that of 1 g of water. This amounts to about 4,800 cm^2. Because of the division of the protoplasmic water into millions of sub-volumes, the total surface area is greatly increased. This is the result of the principle of similitude. The great proliferation of nerve endings thus multiplies the effective surface area by a factor of the order of 1,000. Consequently, the effectiveness of each nerve ending is multiplied. The sensitivity of the human ear toward the oscillatory movements of the insect wing is increased at both the transmitting and receptor ends of the communication system. The principle of similitude increases the ratio of surface to volume in the nerve endings of the ear as well as in the muscle fibers of the insect.

Because of the enormous number of nerve endings in the sense organs, they are adapted to perceive sensations of a small order of magnitude. The rates of oscillation in the 'dielectric valve' correspond to frequencies of the sound waves that are received, but there is a great disparity between the energy required to produce the sound and the energy spent in the act of perception. This should be obvious. The sounds produced by a drummer in a military band require only a small amount of energy originating in his muscular movements. This energy can be propagated over great distances, and can be perceived in the organs of Corti of thousands of listeners. This is explained by the ratio of sensory to motor neurones.

Chapter 10

Communication and Speech

At the present time, the term 'communication' has acquired several kinds of connotation brought about by late developments of the industrial revolution. The term is now used to denote exchanges of information not only among individuals or groups of individuals, but also reciprocal exchanges between man and machines. With the accelerating rates of industrial development, machines have acquired ever expanding kinds of communicative function. They are now used to convey information to other machines or groups of machines, to codify and store human information, and to perform many difficult mathematical calculations that, without their aid, would be almost impossible. The development of calculating machines of many kinds has given rise to the new science of 'cybernetics' (*Wiener,* 1948). This may be regarded as a generalized theory of information and communication, which has found applications to various fields of neurophysiology and studies of the brain.

Consequently, the study of human speech can be approached in several different ways. These would include the methods of physical and physiological acoustics (*von Helmholtz,* 1862; *Miller,* 1916; *Paget,* 1963).

This would embrace the mechanisms of pronunciation of words, vowels and consonants, and the physiology of the organs of sound production and vocalization. It would also involve the neuromuscular coordination of speech with the auditory receptor organs and the exchange of acoustical signals between two or more human beings. This depends on the resonance of oscillatory dielectric valves operating at audio frequencies in both the receptor and transmission ends of the communication system. It also depends on the transmission of sensory and motor impulses through the brain and central nervous systems of two different individuals who are engaged in communication.

Most human speech is audible, and is directed to an audience of one or more individuals at the receptor end. However, certain individuals such as Hamlet and Othello have pronounced great soliloquies; the innermost thoughts of these heroes have been given dramatic form. It is quite possible that many human beings, unknown to Shakespeare, have been given to this form of expression. In such cases verbal expression requires an internal communication system that involves the mind, the central nervous system, the organs of speech, and the operation of billions of sensory and motor neurones and synapses. In the language of modern cybernetics, such a soliloquy would also require an entire system of 'feedback mechanisms' to control speech, gestures and facial expressions. Verbal communication as practiced by a skilled orator or actor involves much more than simple articulation of words and sentences. Silence is also a part of eloquent expression. It would be controlled in part by 'negative feedback' in the central nervous system.

Production of Speech

The main sources of energy in the production of speech are the two lungs. Except in the case of very loud shouting or other unusual conditions of vocalization, the muscular energy requirements are small. These are met by the movements of the flexible rib cage of the thorax and the diaphragm. In normal respiration, air of the lungs is contained in a system of branching tubes carrying two separate masses of air cells that exchange alveolar air with the external air through the bronchioles, windpipe and nasal passages of the upper respiratory tract. The lung itself has no acoustical functions as a resonator. It does not, as commonly thought, produce the 'chest tones' of speech or song. Its function is rather to produce the requisite flow of air at the required pressure and at the right time in coordination with the vocal cords and the organs of articulation.

The two main systems of branching tubes meet in the wind pipe or trachea. In the adult human being, this is an air tube about 2 cm in diameter and 20 cm in length. At the upper end the trachea leads to the larynx. The entire system is flexible, but is supported at the circumference by a strong cartilaginous structure. At the larynx the shape of the windpipe changes from that of a cylindrical structure to one of a flattened form. This leads to the vocal cords, which are contained in a muscular structure shaped like the lips of the mouth, but considerably smaller. These muscular structures lie across the upper flattened part of the trachea and contain the vocal cords. The pitch of the voice is determined by the tension of the muscular tissue, which determines the frequency of the vibrations of the voice. The pitch ranges from one corresponding to a frequency of about 60/sec (low bass) to one of about 1,300/sec (high soprano).

The size and movements of the vocal cords are quite adaptable, depending on the type of speech or musical tone that is being produced. In normal speechless respiration, the muscular lips are spread apart, but in speech or song, the action and positions are quite flexible, and are controlled by neuromuscular mechanisms coordinated by the brain and central nervous system. According to the results that have been described in a number of other chapters, tension and length of the fibers would be related to the intrafibrillar states of water. Thus the frequency of muscial tones produced in the vocal cords corresponds to the tension or magnitude of dielectric energy in the motor nerve endings and in the myofibrils of the neuromuscular system. As explained in chapter 9, the frequency of vibration depends on the neuromuscular properties of the transmitting part of a communication system.

The frequency of oscillation of respiratory energy, as controlled by the tension of the vocal cords, is in resonance with the oscillatory changes of dielectric energy in the auditory nerves of the receptors. The energy requirements of transmission and reception could be of greatly different orders of magnitude. Thus a speaker using unamplified speech of ordinary intensity can be heard by many auditors in a lecture hall or auditorium. The reason for this lies in the great proliferation of sensory nerve endings in the human ear, which enables the auditory nerve to convert sound energy to dielectric energy at a rate of less than 10^{-6} cal/sec. The conversion of dielectric energy to sound energy in the range of audible frequencies may exceed this by a factor of 10^5, as in the wing motion of a small insect (chapter 9). At low intensities the human vocal cords produce sound of a similar order of magnitude to that of the insect wing.

Vocal communication within small groups of human beings depends on anatomical and morphological features of the transmitting apparatus (vocal cords and organs of articulation) and on those of the organ of Corti. The sensitivity of auditory perception is high, due to the great proliferation of nerve endings in the basilar membrane. Vocal communication thus requires only a small expenditure of neuromuscular or dielectric energy in the respiratory passages of the speaker. This energy is mostly spent in the lungs rather than in the neurones and synapses of the speech centers.

Vocal Resonance

In any language, vocal sounds are classified as vowels and consonants, with a number of subdivisions, such as diphthongs. Two main classifications of vocal sounds may be recognized — whisperings and phonations. Thus consonants may be voiced, and accompanied by a hum in the larynx or vocal cords, or unvoiced, that is to say, whispered or breathed. For example, *th,* as in that, and *zh,* as in pleasure, are voiced, while *th,* as in thorax and *sh,* as in shop, are unvoiced. However, all consonants may be recognized when whispered. Each of the consonants is articulated and formed in a particular part of the mouth, with a characteristic method of articulation.

In most languages, the prevalent form of speech is whispered, with the vocal cords being used in an auxiliary sense to give an expressive, emotional or musical quality of the speech. Speech of different individuals varies greatly due to this use of the vocal cords, and by the characteristic rhythms and inflections of each speaker. A number of interesting experiments on the functions of the breath in whispered speech and the phonation of the vocal cords in the auxiliary production of the vowels or voiced consonants have been described by *Paget* (1963). English speech and that of the various other European languages are characterized by their own special modes of breath production and phonation.

In special studies of the vowel sounds of English, *Paget* has been able to identify the upper and lower resonances of 14 vowel sounds. The frequencies of the two resonances extend over a range of about three octaves. These limits are found in the production of the long *e,* as in 'eat'. At the lower resonance, the frequency of vibrations ranges from 304 to 362 vibrations/sec, or from d sharp to f on the musical scale. The upper resonance extends from a frequency of 2,298 (d‴) to 2,579 (e‴) on the scale. Corresponding pairs of resonances are described for the following vocal sounds: it, hay, men, hat, earth, sofa, up, calm, not, all, no, who, put. The upper and lower limits of the two resonances are found in the pronunciation of 'eat'; these extend from d# (304) to e‴ (2,579). These represent the upper and lower limits of the frequencies. A much smaller range of frequencies is exhibited in the pronunciation of 'calm'. In this vowel, the lower resonance occurs at f# (362) and the upper at d#″ (1,217). In the pronunciation of 'who', the two resonances occur, respectively, at f# (9,362) and at d#′ (608).

In these experiments *Paget* found two continuous series of notes, one ranging from about 608 to 2,579, the other from 304 to 912 vibrations/sec. The two series representing the upper and lower resonances of the various vowel sounds were practically independent of each other. The series of notes were due to the resonance properties of the vocal cavities, and are therefore anatomical in nature. They correspond to the overtones and harmonics of fundamental notes produced at lower frequencies in the vocal cords. The methods of identifying and describing each series of resonances is described by *Paget.*

Theoretically, the vibrations in the vocal cords could depend on oscillatory changes of the dielectric properties of intrafibrillar water in the muscular tissue. This produces the frequency of the fundamental not characteristic of the particular vowel sound. Resonant frequencies are then projected by the vocal cavities; these also represent oscillations of dielectric energy in the transmitting apparatus of the vocal cords and cavities. In human speech, the sound is projected as a complex series of vowels, consonants, diphthongs, whisperings and noises, which are analyzed by the receptor apparatus of the organs of Corti of all the auditors. These organs resonate at corresponding frequencies with the vocal cavities of the human speaker or singer.

Communication is thus an exchange of oscillatory energy originating in the central nervous system at the transmission end, and terminating in the auditory nerve endings at the receptor end. At each end there are oscillations of protoplasmic dielectric energy corresponding to fundamental frequencies, overtones and harmonics. Due to the great proliferation of sensory nerve endings in the organ of Corti, sound waves enter the ear carrying much less wave energy than that necessary to propagate them in the vocal cords and cavities. Frequencies and wave forms remain unaltered, but there may be a great disparity in the quantity of energy required to propagate the sound and the dielectric energy that reaches the sensory centers of the auditor.

The Nature of Vowels

Vowel sounds occur in all languages, and have certain common characteristics. They can be continuously intoned, separated from the noises from which they are made to form words. They are classified in different ways. *Von Helmholtz* listed only seven vowels, but *Paget* identifies 14 distinct vowel sounds, as described in the preceding section. A vowel has been defined as 'one of the openest, most resonant and continuable sounds uttered by the voice in the process of speaking'.

The study of the vowel sounds has attracted the attention of many investigators with different points of view. The list includes physicists, physiologists, philologists, vocalists, and many others. It could be extended to musical composers, especially in opera, oratorio and other vocal forms, as well as to librettists and translators. Each of the vowel sounds lends itself to musical expression, because each is characterized by especially favorable pitches and resonances characteristic of the various voices in different parts of the scale. The problem of translating opera into foreign languages gives rise to notorious difficulties. It is the vowel sounds rather than the consonants which are sustained for many beats in vocal music. The great problem in translation arises in having the right vowel in the right place at the right time. Otherwise great masterpieces may become unsingable. The problem arises, for example, in translating *Mozart*'s operas from Italian to German.

A number of investigators of vowel sounds have been listed historically by *Miller* (1916). These include *Willis, Wheatstone* and *Grassmann*. These authorities agreed fundamentally that the quality of vowel tones depends basically on the frequency of the pitch. *Donders* (1866) concluded that the cavity of the mouth is tuned to different pitches for the different vowel sounds. *Von Helmholtz* (1862) developed a 'fixed pitch' theory according to which each vowel is characterized by a fixed range of resonances. This is independent of the fundamental pitch. As applied to the double resonance of a series of 14 vowel sounds, this theory has been elaborated by *Paget* (1963). Some of his results have been given in the preceding section.

Other writers, dating from the period of *von Helmholtz*, have believed that the quality of a vowel is characterized by a series of overtones that accompany a given fundamental. This theory is basically the same as that which determines the quality of tone of certain musical instruments. The overtones depend on the pitch of the fundamental, but the ratios are constant. This is the 'relative pitch theory' of *von Helmholtz.*

Part of the difficulty in arriving at a definite theory of the vowel sounds of human speech undoubtedly arises from the problem of biological variability and reproducibility. No two subjects would be identical with respect to vocal cords and cavities, nor would their habits of producing the vowels be identical. For the present, it would be well to adhere to *Paget*'s theory of double resonance, which had its historic roots in the theories of *Donders* and of *von Helmholtz.* The main point is that the quality of the vowel sounds depends essentially on the anatomical structure of the oral cavity. This would probably be subject to considerable biological variability. In the resting state, the structure and shape of the vocal cavity would be constant for each individual. But in the articulation of words and sentences, the shape and form of the cavity would be in a continual state of flux, altering itself to form each of about 14 vowel sounds. At the same time the phonation of the vocal cords and the intensity of the sound are modified to meet the requirements of expression and meaning.

Physical Resonators

Attempts to produce the vowel sounds of human speech by means of physical models have been made by many investigators including *von Helmholtz, Koenig* and *Miller.* The last-named author has described his vowel-producing system of four organ pipes, and has reproduced photographs of earlier resonators.

Von Helmholtz (1862) designed a system of spherical resonators to assist the ear in the detection of partial tones (harmonics and overtones), as produced by various vowels. Producing a fundamental pitch of 64 vibrations/sec, he was able to detect a series of 19 overtones. The first ten odd-numbered resonators corresponded to frequencies beginning at 128; this is the octave of 64. The spherical resonator is a shell of metal or glass, with one aperture to be inserted in the ear. A second aperture admits the vibrating column of air into the hollow shell, which selects one of the characteristic frequencies. By means of the series of resonators, *von Helmholtz* was able to identify the fundamental frequencies and the various overtones of the vowels that occur in human speech.

About a century before *von Helmholtz, Benjamin Franklin* invented a musical instrument, the glass harmonica, which consisted of a large number of glass drinking vessels partially filled with liquid. Depending on the quantity of liquid and the volume of air in the open vessel, each glass could be tuned to a definite musical pitch. Compositions for the glass harmonica have been written by composers of the period, such as *Mozart,* who was a contemparary of *Franklin.* The glass harmonica produces a characteristic tone that is different from that of any other musical resonator such as a string or percussion instrument.

Since the 19th century a number of methods have been developed for photographing the sounds produced in human speech and in song. The characteristics of the various vowels can then be quantitatively analyzed with respect to frequencies and intensities of the various fundamentals and overtones (*Miller*, 1916). It has been concluded that intoned vowels are periodic and musical. A prolonged musical tone produced by the human voice is not usually sustained for any great length of time. It varies in intensity, pitch and quality. Prolongation of a note would require an absolutely constant flow of breath, constant tension of the vocal cords, and constant resonant properties of the oral cavity. This requires a degree of neuromuscular control that is not given to many singers.

Attempts to duplicate the vowel sounds of human speech by physical models have required the most ingenious studies and methods of physicists who have specialized in acoustics. *Koenig* (1884) has described systems of tuning forks and spherical resonators, which by emitting the calculated fundamental frequencies and overtones, can duplicate some of the vowel sounds of speech. On the same principle, *Miller* (1916) has described systems of organ pipes, four in number, which can reproduce the various vowel sounds. Illustrations of such systems have been published by *Miller* (1916) and by *Paget* (1963). Much earlier it had been observed by *von Helmholtz* that when the loud pedal of a piano was pressed down many of the voewel sounds sung into the instrument could be compared with the groups of strings that resonated with the fundamentals and overtones studied in his experiments.

Consonants

The consonants may be subdivided into various classes which include the semi-vowels and plosives. The semi-vowels comprise *m, n, g* (as in *ing*), *l, r, w* and *y*. Although their linguistic development has been studied by authorities on speech and phonetics, they have generally not been understood as well as the vowels from the point of view of acoustics. In general, the production of the consonants by the tongue and lips is easily understood by anyone who carefully observes his own habits of speech.

In addition to the semi-vowels, the family of consonants includes the plosives *p, b; t, d; k, g.* The fricatives include *s, z; f, v; th* (in thin), *th* (in this). Unlike the semi-vowels, the plosives and fricatives are usually considered to be unmusical noises; aperiodic or unrhythmic vibrations. Photographic records of these sounds cannot be analyzed into families of symmetrical sine waves and harmonics. Unlike those of vowels and semi-vowels, the records are asymmetrical and aperiodic. Thus speech consists of groups of words of multiple tones and resonances of great complexity in which musical tones are flowing and mixed with unmusical noises. Articulation of speech depends on well-integrated neuro-muscular control of the resonant cavities of the mouth coordinated with varying tensions of the vocal cords and with intricate movements of the lips, tongue and teeth. It also depends on well-controlled rhythms and on the application of punctuations and pauses.

Earlier investigators of vocal sounds, such as *Willis* and *Wheatstone* em-ployed physical models such as tubes, cylinders and bell-shaped resonators. Various consonants could be produced by total or partial closure and release of a given resonator. Everyone is familiar with the everyday production of plosives in such operations as the removal of a cork from a wine bottle, with the emission of a distinct 'pop'.

A common article of clothing and of many common household articles, the 'zipper' derives its name from the combination of a fricative, *z*, with a plosive *p*. A 'buzzer' depends for its effect on the conjunction of the plosive *b* with the fricative *z*. The sound of a bee's flight is described as a 'buzz'. The combination of various consonants can be evocative of many natural sounds. The language is full of such descriptive combinations: drips, swishes, splashes, gurgles, whirrings, or howls. One can easily identify the natural consonant sounds in the chugging or sputtering of the motor, the growl or bark of the dog, and the meow or purr of the cat. To a large extent, human speech is based on the faculty of mimicry of natural sounds, as they are produced by animals or by familiar natural objects. These speak to us by recognizable vowels and consonants. Resonant frequencies are produced in the sensory endings of the human cochlea and are transmitted through the nervous system by oscillations of dielectric energy.

In this way, the human body and mind are in direct communication with the inanimate objects that emit natural sounds. Thus the nervous system reso-nates with the howl of the wind, the flow of water and air, and with the continual hum of distant vibrations. The surrounding world continually rever-berates with periodic and aperiodic vowel sounds and consonants, many of which are imitated in human speech, prose and poetry. We rarely experience a moment of absolute silence. Prolongation of such a period would be almost intolerable for most people.

Energetics of Vocal Communication

The energy requirements of various forms of physical and mental activities under various conditions have been compared for adult males (*Morehouse and Miller,* 1967). In such forms of vocal communication as conversation and singing, the consumption of metabolic energy is quite moderate. Taking the

standard basal metabolism for a given individual as 80 kcal/h, the energy requirements for conversation and singing were estimated, respectively, as 110 and
120 kcal/h. The values in excess of the basal levels, are, respectively, 30 and
40 kcal/h. Most of this excess energy is spent in the production of breath by the
thorax and the diaphragm. Comparatively small quantities are expended in the
neuromuscular requirements of the tongue, lips and jaws.

The pitch of musical tones, vowels and the semi-vowels of speech – *m, n, r,
y* and so forth, is controlled by the tension of the vocal cords. This is regulated
by the so-called 'speech centers' of the brain, through which sensory and motor
impulses are organized. The vocal cords function as a pair of transverse lips,
between which expired air from the lungs is passed. In speechless breathing, the
lips remain widely separated and are motionless. In speech, the aperture is
narrowed, and phonation, pitch and quality are controlled by muscular tension
in the lips. These function in a similar manner to the pair of reeds in such
woodwind instruments as the oboe and the bassoon. The tone of these instruments is produced by a longitudinally vibrating column of air produced by the
breath. This determines the energy required to project the sound of the instrument. The pitch and quality of the tone are determined, as in the other
woodwind instruments, by aperatures in the bore of the instrument which may
be closed or opened by the fingers of the performer.

The phonation of vowels and musical tones in the vocal cords depends on
the tension and position of the pair of transverse lips. These function in a
manner that is analogous to the lips of an oboist, a trumpet player or a bugler. In
the vocal cords, the position of the lips may be completely open, nearly
completely closed, or in continuously variable intermediate positions. To produce a given fundamental frequency, it is required to maintain a corresponding
tension of the vocal cords. This can be varied to produce any musical tone
within the range of the singing voice of the vocalist. The range of the bass voice
is from about 54 to an upper frequency of about 724 vibrations/sec, a range of
about 4 octaves (*Paget,* 1963). Compared to the range of about four octaves for
the bass voice, the range of the soprano covers about two octaves, or from about
328 to 1,312 vibrations/sec. These notes range from about e to e″. *Paget* found
for his own voice that pitch and range depended on the time of day. The range
was greatest in the morning, but was diminished in the evening. Pitch and range
differ among the male voices – bass, baritone and tenor, and also between the
voices of women – soparano or alto.

The production of the musical scale by any voice requires a delicately
graded series of muscular tensions in the vocal cords. There would be a one-to-
one correspondence between pitch and isometric tension. Any given isometric
tension is determined by the dielectric energy of the muscle fibers or by the
physicochemical or thermodynamic state of the intrafibrillar water. As in chapter 6, the relation between tension and dielectric energy is given by the formula:

$$-t = 8.60\, c_{Na}{}'' \,\Delta\mu_{Na}{}^0$$

where the dielectric energy is denoted by $c_{Na}{}'' \,\Delta\mu_{Na}{}^0$.

Therefore, frequency and tension depend on the dielectric energy in the vocal cords. In this case, unlike that of the vibrating wings of the bee, the dielectric energy is a function of isometric tension rather than of oscillatory changes of state. The production of phonation in human or mammalian vocal cords is of a different nature from the production of auditory frequencies in the motion of an insect's wings. The source of energy for human vocalization is in the muscular control of breath production. Pitch depends on isometric muscular tension in the vocal cords, and quality depends on the resonating properties of the vocal cords and cavities. Frequencies of musical tones produced by the vocal cords are directly related to isometric tension. This is proportional to the dielectric energy of the intramuscular water, which depends on its state of aggregation or on its dielectric constant. These are colligative properties that are related to the solvent properties of water in any given state. Thus vocalization, speech and communication are essentially colligative properties that vary with muscular tension in the vocal cavities and vocal cords. This is also true of the corresponding changes of state of protoplasmic water and of its colligative or dielectric properties in the nerve endings of the ear. These properties depend on resonant oscillations in the sensory endings. Dielectric energy is then propagated by means of sensory fibers and synapses in the brain and central nervous system of the auditor. Communication is thus a matter of a one-to-one correspondence or mirror image relationship between the oscillating colligative properties of two organisms.

In isometric muscular contraction, the fibers perform no external work. Thus the configurational entropy of a constant isometric tension remains constant, and the value of $T\,\Delta S_c$ is zero. This is the condition for an invariant steady state of homeostasis. In this steady state, the condition of invariance requires that $\Delta G = \Delta H$. Maintenance of a state of high isometric tension requires that the increased rate of glycolysis or free energy production results in a corresponding rate of heat production, as measured by ΔH. Thus there is an increase in the rate of glycolysis but no production of external work in the process of isometric phonation.

In the communication of speech from a speaker A to an auditor B, sound travels at a speed of 1,100 ft/sec from A to B. The longitudinal waves are periodic and aperiodic compressions and decompressions of air that correspond in form to the resonances emitted from the vocal cavities as articulated in the mouth. These waves enter the organ of Corti and the cochlea, where they set up resonant vibrations in the basilar membrane. By means of lateral vibrations, they are communicated to the hair cells and to millions of auditory nerve endings.

By application of the principle of similitude, it may be estimated that the surface area of the protoplasmic water of the nerve endings is multiplied by a factor of several thousand when compared to a continuous spherical mass of the same volume (about 7.2×10^{-8} g of water per nerve ending multiplied by about 10^6 nerve endings).

The dielectric energy of the protoplasmic gray matter is estimated as about 0.05 cal/g of water. This amounts to about 10^{-6} cal/sec as the rate of conversion of acoustic energy that enters the ear to dielectric energy by oscillations of the nerve endings.

Energy conversion involves a process of oscillatory changes of state of the protoplasmic energy from a well-ordered state of low dielectric constant to a disordered state of high dielectric constant. Maintenance of a state of homeostasis or of constant irritability and tone requires a metabolic supply of 'negative entropy' originating in the process of glycolysis. The conversion of glucose in respiratory metabolism must occur at the same rate as the conversion of mechanical (acoustical) energy to dielectric energy in the auditory nerve endings. All these rates would correspond to a value of 10^{-6} cal/sec, taking that rate as the energy of sound reception. Thus constant irritability is sustained in the brain and nervous system by processes of respiration and metabolism that maintain the standard state of protoplasm (water, electrolytes and polyelectrolytes). The general condition for homeostasis requires that $\Delta G = \Delta H$, or that $T \Delta S = 0$. The assumed rate of respiratory metabolism (10^{-6} cal/sec) corresponds to the rate of transmission of auditory impulses, and is extremely low. It amounts to a figure of the order of less than 1 cal/day.

The frequencies of the fundamentals and overtones that produce dielectric oscillations in the nerve endings correspond to those emitted by the vocal cords and cavities of the speaker A. The frequencies also correspond to the tensions that are continually fluctuating in the production of musical tones in the vocal cords. These tones, in the production of speech, are mingled with the various kinds of consonants and noises produced elsewhere in the articulation of lips and tongue. All these sounds are ultimately produced by complex fluctuations of dielectric energy and dielectric constant, as they occur in the organs of speech. These oscillations are then communicated to the cochlea and basilar membrane of the auditor B, where they are converted to oscillatory changes of dielectric energy in the auditory nerve. The process can be represented as:

speech → sound waves → sensory perception
dielectric energy → sound waves → dielectric energy

Thus vocal communication implies periodic oscillations of intramuscular dielectric energy in the sending apparatus transferred by means of sound waves to resonant oscillations of dielectric energy in the protoplasm of the receptor nerve endings. All protoplasm is potentially in mutual relationships with all other protoplasm by means of sensory communication and reception.

As has been mentioned, the human vocal cords produce sound waves at frequencies ranging from 50/sec (low bass) to 1,300/sec (high soprano). These represent 'relaxation times' ranging from 0.05 to 8×10^{-4} sec. The sensory endings of the ear can perceive frequencies as high as 20,000/sec. This corresponds to a relaxation time of 5×10^{-5} sec. In aqueous solution, the relaxation times are of the order of 10^{-7} to 10^{-8} sec (*Oncley,* in *Cohn and Edsall,* 1943). Thus the oscillations in protein solutions are of the order of 10^7 or 10^8/sec, as compared with audio frequencies of the order of 10^4/sec.

Integration of Neuromuscular Behavior

Skeletal muscle is organized as a system of myofibrils and membranous connective tissue. The water exists in several coexistent states: intrafibrillar, intermuscular and membranous. Any given structure exists in a resting or relaxed state and in a stimulated or contracted state. In the resting state, fibers are normally stretched. The water is highly ordered, with a low dielectric constant, behaving as a solvent for fats and other lipids. Due to a low rate of glycolysis, the respiratory quotient (RQ) is low. In this state lipids are transported to intrafibrillary phases, where they are available for mitochondrial oxidation.

In the state of low dielectric constant, the dielectric energy and configurational free energy are high, and configurational entropy is low. Contraction is attended by an increase of dielectric constant, an increase of glycolytic metabolism and anaerobic heat, and either production of external work or an increase of isometric tension. There are also concomitant changes of electrical potential, chemical potentials and standard chemical potentials of the ions. Except for very rapid changes of the chemical potential of water, this function remains constant after the initial process of contraction (*Joseph,* 1971a, 1973).

When one set of muscle fibers A is contracted, another antagonistic set B may be stretched. In that state the dielectric constant of set A is high, glycolytic metabolism and RQ are high, and mitochondrial oxidation of lipids is suppressed. At the same time in the stretched state of B intrafibrillar water is well-ordered, with a low dielectric constant, high configurational free energy and dielectric energy, high standard chemical potentials of the ions, and a high rate of mitochondrial oxidation. The two antagonistic sets of muscle fibers, either in the resting state or in states of muscular activity, may continually transfer energy from one to the other, as each passes from the stretched to the contracted state.

In a state of basal metabolism, this transfer of dielectric energy may occur under conditions of constant total energy, with the entire system continually redistributing energy under fixed conditions of constraint. However, in states of muscular activity that are productive of external work, the movements of antagonistic sets of muscles would be purposefully integrated by the overall behavior. This involves the integrative action of the central nervous system as well as the integrated behavior of all sets of synergistic and antagonistic muscles.

From this point of view, it would be necessary to conceive well-ordered neuromuscular behavior to depend on several kinds of transfer of dielectric energy. These would include intrafibrillar, interfibrillar and intermuscular changes of tension and length, as well as coordinated stresses and strains in the various connective tissues. There would also be transport of dielectric energy at the neuromuscular synapses. These exchanges of dielectric energy would include several subdivisions: voluntary, as in central control of muscular activity; reflex, as in the arcs of spinal segments, and sensory, as in the afferent fibers of various kinds of reflex arcs. All this neuromuscular activity, either voluntary, subject to central controls, or reflex, results in complex processes which determine conduction of impulses in the final common paths of motor neurones.

Everywhere in the neuromuscular system, behavior depends on the integrated conduction of dielectric energy through systems of neurones, synapses and myofibrils. At every point in the complex coherent system, physiological behavior depends primarily on changes of the dielectric properties of the protoplasmic water. These changes of dielectric energy are attended ,by changes of respiratory metabolism at submicroscopic levels. Everywhere in the system, respiratory metabolism depends on physicochemical state. This in turn depends on physiological behavior, which is integrated with the behavior of the whole human being or mammalian organism. At the human level, the integration of neuromuscular behavior with the life of the whole person is well illustrated by the activities of communication and speech. These activities occupy most of our external behavior from infancy to old age. Perhaps for that reason, the physiological basis of biological behavior has been taken for granted rather than subjected to systematic and critical study.

Cybernetics

In recent years, engineers, physicists and mathematicians have collaborated in the development of a new science of cybernetics, which deals with generalized problems of communication and information (*Wiener*, 1948). Within the past 25 years these studies have led to great advances in the development of calculating machines or computers, in which electrical circuits that employ vacuum tubes are employed. By copying the operations of the human brain, engineers have designed the principles of their computers, from the operation of which neurologists and other students of the human brain have learned some fundamental principles of their science (*McCulloch and Pfeiffer*, 1949).

In comparing the operations of computers with those of the brain, one is impressed by the great efficiency of the brain with respect to its energy requirements. The human brain has been estimated to contain about 10 billion nerve cells. The weight of an adult brain is about 1,300 g and the water content is about 1 kg. Therefore there may be as many as 10 million neurones per gram of water, or 10^{-7} g of water per neurone. It has been estimated in chapter 9 that in a given nerve ending in the cochlea of the ear, the water content is about 7×10^{-8} g. Two such endings would therefore have a water content of 1.4×10^{-7} g, which is about the content of one neurone calculated from the figure of 1 kg water for 10 billion neurones. Estimating the dielectric energy of neurones to be of the order of 50 cal/kg water, the energy of oscillation of a single neurone would amount to about $0.050 \times 1.4 \times 10^{-7}$ or 7×10^{-9} cal. For a frequency of 280 oscillations/sec, the energy necessary to operate a 'dielectric valve' would be about 10^{-6} cal/ending per second. If any given nerve ending resonates with only one definite frequency, the number of specific frequencies for the piano keyboard would be 88, assuming gradations of no less than one half tone. But the human ear responds to frequencies of a least two additional oxtaves, or 24 additional notes; this comprises a total of at least nine octaves. It is also sensitive to intervals of less than half tones, and it can distinguish C sharp from D flat.

Thus it might be possible to distinguish at least 200 specific resonant frequencies in the auditory nerve endings. If each ending contains about 10^{-7} g of water, this would correspond to about 50 mg for the protoplasmic water, distributed over 200 different frequencies. The dielectric energy (10^{-6} cal distributed over 200 different oscillators amounts to 5×10^{-9} cal per ending per second.

This figure is of the order of magnitude of 0.050 cal/g of protoplasmic water multiplied by 1.4×10^{-7} g for each nerve ending, yielding a value of 7×10^{-9} cal for the oscillatory dielectric energy of each nerve ending. An ending that is responsive to sound waves of frequencies of 50/sec may transmit 10^{-8} cal/sec (2×10^{-10} cal per oscillation). As pointed out in the preceding section, increased sensitivity depends on the proliferation of auditory nerve endings, which would increase the number of endings that resonate with any given frequency.

It is evident that the dielectric energy that is transmitted in auditory perception is of a very low order of magnitude. It has been pointed out by many authors that, in general, the energy production of the brain and central nervous system in the performance of neuromuscular and sensory functions is very small in comparison with the respiratory metabolism of the organism as a whole. Thus the auditory nerve requires only a few calories per day to maintain homeostasis and normal irritability, as compared with the energy requirement of about 2,000 kcal/day in basal metabolism. For the organism as a whole, neuromuscular respiratory energy is spent in the muscles rather than in the nerves.

Mechanical and electronic computers are much less efficient in their operation than the human brain. In difficult mathematical operations, they are programmed to make instantaneous calculations that are almost impossible for the best brains to perform. However, the brain is very efficient, and with little expenditure of energy, can perform functions of which nonreasoning mechanisms are incapable. The energy requirements of the electronic computers have been described in the following way by *McCulloch and Pfeiffer* (1949). 'A computer with as many vacuum tubes as a man has neurones in his head would require the Pentagon to house it, Niagara's power to run it, and Niagara's water to cool it.' Even when vacuum tubes are replaced by transistors, the human brain is more efficient in its energy requirements by a factor of hundreds of thousands. This high efficiency of the brain is brought about by the very low energy requirements of sensory endings, nerve endings and synapses in the brain and nervous system. The number of neurones in the brain is of the order of 10 billions. In addition there are tens of billions of spinal and peripheral neurones, all of which make synaptic junctions with other sensory and motor neurones. The number of such oscillatory relays and circuits in the nervous system may be of an almost inconceivable order of magnitude.

The electronic computers are capable of mechanical operations of very high levels of integration and abstraction. They are capable of storing and retaining information in ways that have little resemblance to any other kind of mechanism aside from the human brain, sense organs, and neuromuscular systems. Like the human brain and body, the mechanisms of the calculators are goal-directed. Electronic computers are 'servomechanisms', in which some of the output of energy is fed back to the source to regulate the output. This 'feedback' in servomechanisms may be used to maintain existing processes in steady states, or it may be continually used to redirect the control of performance. Technological analogies are found in the guidance of a missile, the trajectory of which is directed by radar signals from a moving target. Many examples of negative feedback could be found in human or animal behavior, in which muscular movements are goal-directed or oriented. Examples are found continuously in human speech and conversation, in which the output of one speaker is guided by negative feedback resulting from the remarks, facial expressions and gestures of his auditors. Such behavior is second nature to audiences and speakers who communicate not only verbally, but also by applause, smiles and frowns, and not uncommonly by coughs, boredom and fatigue.

Like the act of auditory communication, these examples of cybernetic feedback, servomechanisms and goal direction are such common examples of everyday experience that their basis in physiology and in the physical chemistry of protoplasmic morphology and behavior may be either ignored or taken for granted. It is generally agreed that verbal and auditory communication is of human value. It follows that physiology and biological science cannot long ignore questions of value and purposefulness. This is also true of physical chemistry and biological chemistry in their general approach to the study of vital phenomena.

Chapter 11

Ordered Processes

The behavior of living organisms, including that of all animals and plants, is studied in the science of *ethology*. Practically all known species are members of communities or associations belonging to a common biocoenose (*Wheeler,* 1928). Within such a community, the ethology of all members is a property that depends on mutual relationships with all the subgroups including predatory, parasitic and symbiotic forms of behavior. The resulting life of the community is an *emergent* of the morphology and behavior of each living unit or entity. Environmental factors such as climate, topography, temperature, humidity and sunlight also play important roles in the life of any community or association of living things.

Thus the life of any species is related in one way or another to events in the cosmos — the motions of the earth and tides, the seasons of the year and the rising or setting of the sun. In this way, the vital activities of all animals and plants are related not only by mutual ethological factors, but also by all events in the physical universe. These depend on the existence of 'eternal objects' in the external world, which tend to maintain states of order and coherence. These provide invariant stabilizing edaphic and biotic conditions. Invariant chemical composition of the atmosphere, of the oceans, lakes and rivers, and of the soils and mineral deposits are factors that provide order and coherence.

Perfect order in the universe cannot be maintained by the presence of the eternal objects. On a geological time scale, there are cosmic or earthly catastrophes of varying degrees of severity. Pollution of the atmosphere and of the waters of the earth are recurrent problems. Erosion of the soil, floods and changes of river beds and water levels are also factors of disorder. Many other terrestrial changes, both favorable and disadvantageous, have operated on both the historical and geological time scale to change the course of ethological evolution. These changes have been largely accidental in prehistoric and paleolithic ages, but since the neolithic ages of agriculture, they have become well-ordered and purposeful. This has been almost entirely a result of human reason operating to control the physical environment and to eliminate natural forces of disorder. This is evident in the development of agriculture, the control of plant life, the cultivation of trees, forests and parks, and in the domestication of animals. All these developments may be regarded as the results of human efforts to control plant and animal morphology and to introduce order and discipline into animal and human ethology. The developments are fundamental functions of reason, which are to develop the art of human life.

Order and Purpose

The term 'order' is relative. It acquires meaning only with reference to correlative states of disorder. In general, the term connotes 'givenness' (*White-head*, 1929). 'There can be no peculiar meaning in the notion of *order* unless this contrast holds ... Each actual entity requires a totality of *givenness*, and each totality of *givenness* attains a measure of order.'

According to these notions of order and disorder, the morphology and physicochemical state of any complex animal organism cannot be defined according to any absolute scale, but can refer only to a state of *givenness* at any time or place, under *given* conditions in its cosmic environment. Absolute states of order or disorder cannot be presumed in either the organism or in the environment. In general, purposefulness in animal morphology and behavior presupposes states of order, but this is also a matter of 'givenness'. Neither chemical morphology, ethological behavior or purposeful goal directedness can be presumed to be absolute invariants. 'Homeostasis' is an invariant physiological and physicochemical state, but it should be regarded as a 'self-identity' or state of 'givenness' rather than in relation to any absolute standard. All physiological and biochemical standards are subject to normal statistical deviations. Perfect order in organism and environment is an ideal theoretical limit, like the perfect gas laws, but is perhaps a necessary and useful approximation in relating purposeful behavior to chemical morphology.

According to the views of *Frey-Wyssling* (1953), cited in chapter 1, well-ordered, *purposeful* biological processes can only be understood in relation to well-ordered submicroscopic morphology of protoplasm in all cells and tissues. This depends on the molecular nature of the genes, which are to be regarded as aperiodic crystals of high molecular weight (*Schroedinger*, 1944). These are centers of low entropy and crystalline order in the cell nuclei, and provide 'information' in the well-ordered synthesis of all intracellular and extracellular macromolecular substances. Because of Brownian movement and other statistical errors due to chance ('tychism'), growth and development of cells and tissues would not be perfectly ordered, except in the ideal limit. However, growth and development are well-ordered, and according to *Carnot*'s principle and the phase rule, proceed in time with 1 degree of freedom. The system is thus univariant and held together by a process of synechistic unity and coherence. According to *Caratheodory*'s statement of *Carnot*'s principle, growth and development are channeled into the most accessible states. All the secondary physiological and physicochemical properties are functions of this normal state of morphology. Fundamentally, they depend on 'information' from the aperiodic structure of the genes.

Within this synechistic development, biological processes are limited by tychism, which supplies an element of freedom. Thus body weight, muscular development and neuromuscular aptitudes are free to develop within limits that are imposed by the genetic and thermodynamic constraints. Such processes depend on nutritional factors, habits of work and exercise, and on individual pleasures and tastes. These factors require freedom of the will, and are both tychistic and agapastic, but they are strongly determined by the synechistic conditions of genetic and thermodynamic continuity. The well-ordered processes of life are thus continuous with morphological development, but not in any absolute one-to-one biunivocal set of correspondences. Factors of chance and accident must be considered; these work against perfect order both in the organism and in the environment.

The milieu intérieur

'The stability of the milieu intérieur is the primary condition for freedom and independence of existence; the mechanism which allows of this is that which ensures in the milieu intérieur the maintenance of all the conditions necessary to the life of the elements. From this we know that there can be no freedom or independence of existence for simple organisms whose constituent parts are in direct contact with their cosmic environment, and that this form of life is, in fact, the exclusive possession of organisms which have attained the highest state of complexity or of organic differentiation' (*Bernard,* 1878).

This definition of the functions of the milieu intérieur in mammals (vie constante) clearly affirms the principle of order and stability in the internal environment. It also implies that this is a necessary condition of reciprocal order in all the cells and tissues. This implies a property of 'fitness'. In *The Fitness of the Environment, Henderson* (1908) has shown that the blood plasma of higher organisms has evolved with the conservation of the inorganic ions that are derived from early forms of sea water. According to *Macallum* (1910, 1926): 'Enough, however, has been advanced here to make it extremely probable that the inorganic composition of the blood plasma of vertebrates is an heirloom of life in the primeval ocean.'

Thus organic evolution has progressed over long periods of geological time with the conservation of characteristic ionic concentration ratios that represent the primitive geochemical distribution of mineral salts in paleozoic ages. This depends on the 'conservation of reversibility' of the vital processes insofar as they are related to electrolyte distribution in the animal organism. As shown in

chapter 6, 'Biological Development', this is shown by such characteristic thermodynamic functions as the standard free energies of ion distribution throughout
the animal kingdom. These functions tend to remain constant in such various
structures as the oocytes of sea urchins, and in skeletal muscle of many
invertebrates and vertebrates, including those of mammals. Thus the evolutionary process maintains certain conditions of invariance in phylogenetic development. These constant conditions are those that are 'necessary to the life of the
elements' (*Bernard*, 1878). Thus order in the higher forms of life is a matter of
reciprocal fitness of the milieu intérieur and all the other tissues of the body.

Order and fitness are colligative properties (*Joseph*, 1973). The chemical
potential of water is reciprocally related to the electrolyte composition of the
milieu intérieur. In mammals, the value can be estimated as about -3.1 cal/
mole, as referred to the chemical potential of pure water at the freezing point.
This value is estimated from the average value of the freezing point depression of
mammalian blood serum. Thus:

$$\mu_{H_2O} - \mu_{H_2O}^0 = 5.26 \, \Delta$$

where Δ is the freezing point of serum, and 5.26 is the entropy of fusion of ice
expressed as calories per mole per degree. When Δ is taken as $-0.58\ ^{\circ}C$, the
value of μ_{H_2O} referred to $\mu_{H_2O}^0$ is -3.05 cal. As an invariant property of the
milieu intérieur, this depends on the reversibility of all processes of distribution
and transport of the physiological ions between the liquid phases and the
intracellular phases in which water and electrolytes are reversibly distributed.
These are phase rule conditions of invariance and constraint which stabilize all
distributions of the components between cells and tissues. According to the
phase rule, these depend on the conditions

$$\mu_{AB}' = \mu_{AB}''$$
$$\mu_{AB_2}' = \mu_{AB_2}''$$
$$\mu_{H_2O}' \leqslant \mu_{H_2O}''$$

These conditions of order, balance and constraint were derived by *Gibbs*
(1875) in his equations 77 and 78. In the case of water, the inequality sign
expresses the fact that metabolic water (produced in cellular respiration) tends
to be transported irreversibly from the cells to the extracellular phase. The rate
of transport is determined by the rate of formation of water, and this in turn
depends on the rate of respiration and of heat production. These represent rates
of irreversible transport that are necessary to maintain the normal state of
'givenness', order and homeostasis. They are derived ultimately from *Carnot*'s
principle, which, as has been explained, maintains the organism in highly accessible
states. Biological processes depend on changes of all the colligative proper-

ties, including the electrical potentials. These are related to changes of state of water, which determine dielectric constant, dielectric energy, metabolic rates and transport, and other secondary properties such as muscular contraction and tension, and nerve impulses.

Givenness

As an example of a well-ordered body fluid of emergent and synechistic evolution, mammalian or human blood plasma is typical. For all species of mammals, plasma has approximately the following ion concentrations (in mEq/l):

Na	K	Ca	Mg	Cl	Total
140	5	5	2.5	110	262.5

Expressing the values with sodium as the basis of reference, the values are as follows:

Na	K	Ca	Mg	Cl
100	3.57	3.57	1.78	78.5

For sea water of the present era, the comparable results are (*Joseph,* 1973):

Na	K	Ca	Mg	Cl
100	2.11	2.23	11.4	117

If the values for mammalian plasma are assumed to represent the composition of a prehistoric state of sea water, it is seen that the ratio of sodium to potassium has increased; this is also true of the ratio of sodium to calcium. On the other hand, the ratios of sodium to magnesium and of sodium to chloride have decreased. Thus the ordered state of water has changed over the geological time scale. The ocean has remained well-ordered for millions of years, but has changed with respect to its 'givenness'. However, at no time could the composition be said to be disordered. The development has been continuous and synechistic. The total electrolyte concentration has increased from about 260 mEq (as represented by the present value for mammalian blood plasma) to one of the order of 1,100 mEq for sea water of the present era. As shown above, the most remarkable increase has been that of magnesium; the present ratio is about 5.7 mole of Mg to 100 mole of Na. The prehistoric ratio was of the order of 0.9 to 100. This required an adaptation in the osmoregulatory mechanism to control blood magnesium in marine invertebrates (*Joseph,* 1973).

The evolution of the mammalian kidney has led to a remarkably accurate control of the electrolyte composition of the blood. This development of the present state of 'givenness' has been attended by concurrent evolutionary processes in all the other systems of homeostatic controls necessary to maintain the given state of order at any period of the phylogenetic process. These include the development of the cardiovascular system, the lungs, the kidney, with simultaneous controls of acid-base balance and respiratory metabolism. The biochemical evolutionary processes involve the development of serum proteins, hemoglobins and other respiratory pigments, of the capillary circulation, and of neural and hormonal control of water and electrolyte balance in the ground substance of connective tissues (*Engel et al.,* 1961; *Gersh and Catchpole,* 1960). Thus order, homeostasis and 'givenness' are emergent and synechistic properties of the total chemical morphology of all the cells and tissues. The development has not been mechanistic, Cartesian or Newtonian. It has been limited to the thermodynamically and genetically determined accessible states, which are organismic and emergent.

As *Henderson* (1908) was the first to show, the development of life in all species depends on the fitness of many of the important physicochemical components of living systems. These include the fundamental properties of water in all its states. These are related to the solubilities and physicochemical properties of the inorganic electrolytes, carbon dioxide, oxygen and to acid-base balance in any system of cells and tissues. These properties are related to the geochemical properties of the environment and to geophysical properties such as temperature and atmospheric pressure. These constitute the conditions of 'givenness' at any place in the biosphere and at any epoch of geological time. These conditions have been subject to slow or continuous development (synechism) at most times, but they have also been subject to catastrophic and discontinuous changes (tychism). The latter types of change can produce effects of disorder both in the environment and in the life of animals and plants. As a catastrophic change of 'givenness', the transition from the age of reptiles to the age of mammals and birds may be cited. This transition occurred during geological periods from the Permian and Triassic (250 million years) to the Jurassic and Carbonaceous (150 million years).

At that period, profound changes in geochemical, geophysical, edaphic and biotic conditions must have occurred, requiring adaptations to the new conditions of order and 'givenness'. At no time in the geological history of the earth have these conditions become intolerable for *all* the existing species. At all periods, synechistic continuity and environmental order and 'givenness' have been maintained.

Changes of Order

Many kinds of cells and tissues, especially of the mammalian neuromuscular system, are characterized by the properties of irritability and contractility. All kinds of biological processes and behavior depend on changes of physisochemical state of this kind. In the contraction of human or mammalian skeletal muscle, for example, the fibers pass from a well-ordered relaxed state of aggregation to a relatively disordered state of contraction. When the contractile process is ordered and purposeful, the limiting states of contraction and relaxation are characterized by definite states of 'givenness'. These are measured by the tension and length of the fibers, by the resting, action and 'spike' potentials of the process, by the maximal work of the contraction, and by the electrical properties of the dispersion medium. Dielectric properties depend on order and disorder of the intrafibrillar aqueous milieu, which determine the dielectric constant, the dielectric energy, standard chemical potentials of the ions, and respiratory metabolism.

Other secondary properties include the electrical potentials, as related to the 'givenness' of the intracellular states of water. These are 'colligative' or 'synechistic' properties, which are 'held together' by the physiological or physicochemical state of the entire system, as defined by the chemical morphology.

Changes of state in the neuromuscular system may be classed in two different ways. The first type of change would include those changes of order and disorder which do not result in production of external work or the change of the total respiratory metabolism. Such changes would apply to distributions and redistributions of muscular tension in various sets of antagonistic and synergistic sets of fibers with no net result in production of external work or changes of the total respiratory metabolism. Such changes would apply to distribution and redistribution of muscular tension in various antagonistic and synergistic sets with no net change of the total configurational entropy of the body. The conditions for such processes at constant configurational free energy, dielectric energy or configurational entropy of the body require:

(A) $\Sigma \Delta G_c = 0$
(B) $\Sigma T \Delta S_c = 0$
(C) $\Sigma \Delta H = \Sigma \Delta G$

According to condition A, free energy may increase or decrease in a given set of fibers, but only under the condition that the total free energy remains constant. This requires compensating changes of tension in all the muscles of the body. Changes of posture, facial expression, and other types of behavior are permitted under conditions of constant configurational free energy and entropy (conditions A and B). The condition for this is given by C, which requires perfect energy balance and constant respiratory metabolism. Thus the first type of behavioral response is *conservative;* it yields no external work or other kind of physicochemical change.

In the second, or *nonconservative* type of behavioral response, there are changes of order and 'givenness' in the body that are related to corresponding changes produced in the environment. These changes are uncompensated in the body, and result in *irreversible* changes in the external world; external work or *irreversible* production of heat and free energy. The conditions for such *nonconservative* processes are given by:

(A) $\Sigma \Delta G_c < 0$
(B) $\Sigma T \Delta S_c > 0$
(C) $\Sigma \Delta G < \Sigma \Delta H$

Conditions A and B show that the internal physiological processes result in uncompensated losses of configurational free energy and increases of configurational entropy. These require the condition C that implies a state of negative free energy balance within the organism. All these changes depend on *increases* of the intramuscular dielectric constant, increases of entropy or disorder and decreases of dielectric energy. Increased rates of glycolysis that accompany muscular contraction also depend on changes of the aqueous dielectric. The state of disorder results in the solubilization of glucose, glycogen, phosphagen and other metabolites. This is the result of a change of dielectric constant from a resting state of about 30, to an active state in which D'' approaches 80.

Depending on the relative states of order and disorder in the neuromuscular system, there are corresponding states of the 'givenness' of the physicochemical state of the body as a whole. Behavior and chemical morphology are thus continuous and synechistic with entire sets of properties. In human life both processes change continuously or discontinuously from instant to instant. They are emergent properties that cannot be reduced to mechanistic or Cartesian components. Thus the classical dualistic concepts that attempt to isolate mind and body from the external world cannot account for the relations between animal morphology and ethology. Neither can they be applied to the separation or isolation of the internal and external aspects of behavioral processes.

'Givenness' is a conservative property on a day-to-day basis. During periods in which the organism operates on a nonconservative basis, there are increases of disorder in the active cells and tissues. These results in periods of negative free energy balance and changes of morphological 'givenness'. However, when the active periods of negative free energy balance are followed by resting periods of positive balance, the initial state of 'givenness' is restored, and the ordered state of homeostasis is regained. This requires a suppression or inhibition of glycolytic metabolism, and a period of mitochondrial oxidation of lipids. In this recovery period, the respiratory quotient is lowered, and configurational free energy and entropy are restored to the well-ordered resting state (*Joseph,* 1973).

Animal Locomotion

The great majority of land mammals are quadrupedal. They include all dogs, cats, and other familiar species such as rodents and rabbits. Hoofed quadrupeds include horses, cattle, deer, elk and many other families of ungulates. In general, each species practices its own special kinds of locomotion, but for any given species, these may fall into different categories. In dogs and foxes, the typical *walk* can be described as 'diagonal' (*Bourlière,* 1953). Two diagonally opposite feet are simultaneously lifted as the other pair remains on the ground to support and propel the animal. The elevated pair is then placed on the ground, while the alternate diagonal pair is lifted. The method of locomotion may be represented as:

$$
\begin{array}{cccccc}
\text{C} & \text{A}' & & \text{C}' & \text{A} & & \text{C} & \text{A}' \\
\text{D}' \quad \text{B} & & \rightarrow & \text{D} \quad \text{B}' & & \rightarrow & \text{D}' \quad \text{B} & \text{etc.}
\end{array}
$$

$$
\quad\quad \text{I} \quad\quad\quad\quad\quad\quad\quad \text{II} \quad\quad\quad\quad\quad\quad\quad \text{III}
$$

As the animal walks in the direction from left to right, it assumes a series of odd-numbered positions (I, III, etc.) that alternate with a series of even-numbered positions (II, IV, etc.). In the odd-numbered positions, the left foreleg and the right hind leg (a diagonal pair) are simultaneously lifted, while the other diagonal pair (B, C) are extended with the feet on the ground. Quadrupedal motion proceeds with each diagonal pair (A, D) and (B, C) alternately lifted and extended. The motion is thus well-ordered and characterized by two alternating states of 'givenness'. The flexor and extensor muscles of the four legs do not operate independently, but are coordinated by the nature of the walk or gait. This is an organistic property of the species, but it depends in various cases on the 'givenness' of external conditions. When the external situation changes, the character of the quadrupedal motion changes as a behavioral adaptation.

In the form of locomotion known as *pacing,* it is the two isolateral legs that are simultaneously advanced, rather than the diagonal pair. This may be represented as:

$$
\begin{array}{ccccccc}
& C & A & & & C' & A' & & & C & A \\
D' & & B' & \rightarrow & D & & B & \rightarrow & D' & & B' & \text{etc.} \\
\end{array}
$$

$$
\quad\quad\quad\text{I}\quad\quad\quad\quad\quad\quad\quad\quad\text{II}\quad\quad\quad\quad\quad\quad\quad\quad\text{III}
$$

Here the odd-numbered positions (I, III, V, etc.) correspond to forward positions of the left legs (A, C) and elevated positions of the other pair (B, D). In the even-numbered positions, A and C are elevated while B and D are extended. All sets of muscles are therefore coordinated and in well-defined states of 'givenness' at all stages of the walk. Thus quadrupedal locomotion can take many different forms, depending on the species and on the specific situation to which the motion is adapted. In all cases, however, the motion is well-ordered and adaptive. It is characterized by well-defined states of 'givenness'.

As in all cases of muscular contraction and extension, there are simultaneous and alternate changes of respiratory metabolism. The well-ordered states of extension are characterized by low entropy and low intracellular dielectric constant. In this state dielectric energy and configurational free energy are high. The muscle fibers are highly polarized and characterized by high standard chemical potentials of the ions. In the alternating state, the situation is reversed. In muscular contraction or flexion, the dielectric constant is increased, anaerobic glycolysis is stimulated, electrical depolarization occurs, and dielectric energy is converted to external work. In quadrupeds, all these processes are coordinated with the external situation. All behavior, then, may be regarded as depending on well-ordered adaptive processes, which involve all the colligative properties of the neuromuscular system. These are states of 'givenness' which cannot be reduced mechanistically to the physicochemical responses of isolated myofibrils, nerve fibers or reflex arcs. All behavior is therefore phylogenetic, ontogenetic and organistic. External adaptive behavior is therefore to be regarded as synechistic with all internal changes of chemical morphology, metabolism and other states of 'givenness'.

From this point of view, external behavior is related to internal morphology and metabolism by one-to-one biunivocal or 'mirror-image' sets of correspondences. The neuromuscular changes are of the nature of emergent 'Gestalten' rather than summations of additive mechanistic processes. According to the principles of 'emergent evolution' (chapter 1), it is impossible to reduce animal behavior to isolated processes in the sense of *identification* (*Spaulding*, 1918).

Contraction of skeletal muscle yields external work at the expense of dielectric energy or configurational free energy at the muscle fibers. This implies an increase of configurational entropy, ΔS_c, related to the external work:

$$
\Delta G_c = -W_{max} = T \, \Delta S_c
$$

Recovery of the configurational free energy or restoration of the configurational entropy requires a return of the fibers to the extended and well-ordered state. This is brought about by 'negative entropy', obtained from anaerobic glycolysis. In the production of 2 mole of lactic acid from 1 mole of glucose:

glucose = 2 lactic acid

the free energy change, ΔG, is -29.88 kcal. The other constants are: $\Delta H = -17.67$ kcal and $T \Delta S = 12.21$ kcal. The increase of entropy is obtained by the formation of 2 mole of a triose from 1 mole of the hexose, with the breaking of the 6-membered pyranose ring structure. Then the repolarization of the stretched fibers involves a state of high order and 'negative entropy'. This is obtained from the metabolic supply of 'negative entropy', maintained in muscle by nutrient sources of glucose and glycogen. The function of ATP is not to supply phosphate bond energy to the contracting fibers, but to react in the formation of glucose-1-phosphate and creatine phosphate ('phosphagen') in anaerobic glycolysis. Formation of various phosphate esters is an integral part of glycolytic respiration. These processes are accelerated in muscular exercise.

The function is to maintain homeostasis, muscular tone and irritability of the stretched fibers by supplying negative entropy from the nutrient sources. Thus the function of ATP is not to supply 'bond energy' to the contracting fibers, but to enter into intermediary steps in glycolysis, thus maintaining the normal homeostatic states of order and negative entropy in the stretched or uncontracted muscle (*Joseph,* 1971a, b; 1973). Thus intracellular respiratory metabolism functions in all behavioral processes to maintain negative entropy, irritability, tone and 'givenness'. Respiratory metabolism is largely nonspecific in relation to ethology and genetic factors.

Upright Posture

The upright posture of man has evolved over a period of millions of years. It has been attended by parallel evolution of all other parts of the body required to conform to the vertical position. In particular there have been fundamental changes in the special form of the human skull as an adaptation to the upright gait. In mammalian quadrupeds, the mouth, jaws and oral cavity are extended in a relatively horizontal position and are situated in an anterior position with respect to the brain. This places the teeth and jaws far in advance of the shoulders and forelegs. The head, nose and mouth serve many functions related to the senses of smell and taste, and in the wild state are positioned to serve many of the normal functions of hunting, aggression, defense and feeding. The legs of a quadruped are mainly adapted to locomotion, to climbing, and to digging. Animals have not developed the faculty of using tools or of using the forelegs in any way but the most primitive kinds of manipulations.

The earliest ancestral forms of the human race began possibly to separate from the apes about 20 million years ago in the age known as the Miocene. Remains dating from a prehistoric period from a half million to 2 million years ago were discovered in eastern Africa by *Dart* in 1923. His findings were later extended by *Broom* and *Robinson* and by *Leakey*.

According to *Darlington* (1969), these early forerunners of man had made several astonishing advances over the ape-like forms of the Miocene.

(1) They made tools from horn, bone and stone.

(2) The teeth had evolved toward the human state.

(3) They had progressed toward the upright position.

(4) However, the brain capacity remained small – about 450–550 cm^3 as compared with a maximum of 750 cm^3 for an ape.

Great variability of the fossil forms from this era has been found. Evidently, thousands of generations of evolution were required to stabilize human anatomy, and especially the size and shape of the human skull to the dimensions now characteristic of various races of man. The detailed nature of the evolution of the human skull in adaptation to upright gait has been studied by *Waldenreich* (1924) and by *Sicher and du Brul* (1970). One of the great changes has been the great evolutionary change of the brain capacity, which is now of the order of 1.3 liters, or 1 kg water.

In great measure, man's evolutionary progress with respect to ethology and behavior depended on the development of the arms, hands and fingers. This enabled him to develop many kinds of tools and weapons, and to attain great superiority over all other species with respect to hunting, defense and security of nutrition and feeding. These early achievements enabled man to abandon the forests and trees as habitats and to live in open spaces or grasslands. Quadrupedal locomotion was abandoned in favor of the characteristic upright gait. These developments preceded the enlargement of the brain capacity, and the more recent evolutionary development of perception, cerebration and great manual dexterity and skills. All such behavioral development would have required continuous and synechistic development of the gross anatomy, held together by well-ordered and purposeful development of submicroscopic morphology of the cells and tissues. The most constant or invariant property would have remained the invariance of the milieu intérieur. Like that of all mammals, the internal environment must have retained the electrolyte and water content characteristic of paleozoic sea water (*Henderson,* 1908; *Macallum,* 1910, 1926). Constant water and electrolyte composition would imply constant chemical potentials of water and electrolytes in all cells and tissues at all stages of phylogenetic and ontogenetic development. These are the constraining conditions of all ordered physicochemical behavioral processes. All such processes depend on changes of state or order determined by the dielectric properties of water in each kind of tissue. These determine configurational entropy, free energy, maximal work and isometric tension. They are also related to neuromuscular controls and sensation and perception.

In the development of manual dexterity and skills, many of the primitive properties of skeletal muscle have been retained. Thus the thermodynamic functions such as standard free energies and standard chemical potentials have remained fairly constant in the skeletal muscle not only of mammals, but also in cold-blooded vertebrates and invertebrates (chapters 5 and 6) (*Joseph,* 1971a, b; 1973). Thus behavior depends largely on changes of gross anatomy rather than on the more primitive properties of protoplasm related to the dielectric properties. According to phase rule reasoning, gross anatomy is extensive and adaptive. It is a first-order thermodynamic function (*Joseph,* 1971a, b; 1973). Chemical potentials and dielectric properties, on the other hand, are zero-order functions and tend to be ontogenetic and phylogenetic invariants. Ethological evolution thus depends largely on changes of the first-order functions, which are highly adaptive and tychistic. The zero-order functions, on the other hand, are mainly synechistic, continuous, colligative and nonadaptive.

Language

Whatever the genetic mechanisms of human evolution may be, there can be no doubt that the developments of the skull that led to the perfection of the human speech apparatus were of the utmost importance in human survival (*du Brul,* 1958). The development of the numerous human languages was a well-ordered purposeful process that was continuous and synechistic with the morphological changes.

The study of the hundreds of languages and dialects now in use in the modern world is of interest not only to scholars in the various fields of linguistics, but also to travelers, men of business and commerce, and also to explorers and anthropologists. A contemporary anthropologist, *Levi-Strauss,* has shown the relation of language to human behavior among many different groups of 'primitive' peoples. In *The Savage Mind* (1970), he has criticized many of the conceptions and preconceptions that have been presumed in previous studies of the subject. For example, he has shown that among the various nations and tribes of the North American Indians, the various languages can be very rich, not only in concrete terms for the individual objects of everyday practical interests, but also in the abstract terms of general or theoretical interest. Thus the so-called primitive mind has no great difficulty in forming the idea of universals as contrasted with the particulars. This was one of the problems that occupied the minds of the philosophers and theologians of the scholastic philosophy of medieval Europe — *Duns Scotus, Peter Abelard* and *William of Occam*. The reality of universals was questioned by many of the medieval nominalists. Theirs was a heretical view in theology and metaphysics. The problem was rooted in the ideas inherited from theological sources, and by the metaphysical doctrines of Aristotelian and Platonic origin. Primitive peoples, with no such metaphysical heritage, appear to have solved without difficulty and in their own way the problem of universals and particulars. Language, to serve both the practical and abstract interests of life, must handle these questions effectively.

Contrary to popular belief, 'primitive' peoples often evince great interest in general or abstract ideas quite apart from any practical or economic interest or application. Often they take pride in their 'poetic' approach to nature and to man, and tend to reproach advanced civilizations for their practical and utilitarian views. To a large extent, these views are a result of centuries of domination by nominalism, which tends to undervalue the abstract, the poetic, the mystical and the universal. Thus modern languages are very rich in the nouns required for a technological society — the innumerable items of commerce and daily life. They are rather impoverished with respect to the ideas and symbols of ancient mythology and of medieval religion.

Thus language at any time or place is perhaps the best reflection we have of human behavior at all levels — practical, social, religious, scientific and philosophical. Language is inseparably bound to the total life of any people at any stage of history. It is now clear that the development of language, depending on the development of the skull, oral cavity, mouth and jaws, was a necessary part of human development from the old stone age or paleolithic period to the neolithic period. Over periods of thousands of years, man obtained increasing control of his environment. This required development of tools of increasing effectiveness, of weapons and hunting equipment, and of various agricultural and domestic arts. These included a long-term process of domestication of animals, and of methods of cultivating the land and control of plant life. Arts and crafts such as pottery making and weaving also would have required thousands of years of continuous and systematic advances from the paleolithic to the neolithic age. Systematic progress of this kind would have required purposeful behavior and long continued development of language and communication. Thus from the point of view of paleontology and archaeological history, the development of language must have been inseparable from that of morphology and of the practical and domestic arts and sciences.

Chapter 12

Morphology and Behavior

Animal and human behavior are in varying degrees constrained by conditions of morphological growth and development. Behavior therefore must depend on principles of chance (tychism) as well as on laws of continuity (synechism). Biological development, including the evolution of all species of animals and plants, is a biological process that implies the operation of the laws of continuity as well as the laws of probability or chance. Without the factors of chance and probability, the development of animal morphology and behavior would be inconceivable. Evolutionary development cannot therefore be applied or be explained by Newtonian or Cartesian principles. On the contrary, it involves the principles of population genetics. In any biological species, the dynamics of population depend on irreversible trends in ecology, as they are related to environmental changes. To a large extent these depend on the evolutionary changes of human behavior.

Far beyond the primitive properties of irritability, sensation and feeling that are inherent in all protoplasm, we must include the vast historical achievements of the human mind in all the realms of its imagination and creative faculties. These realms embrace all the arts, sciences, technologies, and intellectual achievements of a historical development that covers the lives of thousands of generations of the human species. This implies that all these achievements have occurred and will occur within the potentialities of living protoplasm. The laws of continuity impose the constraining conditions of synechism, which require a continuous 'holding together' of morphology and behavior, which cannot develop independently. These developments will continue beyond any foreseeable future. Potentialities will be limited only by constraints imposed by the laws of probability or chance (tychism). The laws of development depend on constraining conditions of reversibility (continuity) and on those of irreversibility (discontinuity).

Presumably it will be impossible for humanity or for any other species to evolve new potentialities that depend on different kinds of protoplasm or genes. New kinds of atmospheric gases or mutations of the chemical elements cannot be presumed. Future evolutionary changes will therefore be limited to the principles of synechism, which imply the endurance of physicochemical and biological conditions of constraint. New kinds of protoplasm or genes would imply mutations of a kind unknown to us. They are not necessarily inconceivable, but they are incompatible with the principle of continuity or synechism. They would be due to the operation of chance (tychism) which, unlike synechism, permits irreversible discontinuous breaks in the historical or evolutionary process. Defiance of the principle of continuity would imply a catastrophe of cosmic dimensions. Transformations of the chemical elements, of the atmosphere, and of the earth itself would be involved. The existence of eternal objects could no longer be presumed. In the course of such a catastrophe, the emergence of new forms of life would require the evolution of new kinds of living substances or protoplasm. It would not be permissible to conceive some kind of vaporous state of life now imagined only by spiritualists or other kinds of supernaturalists. For the present, we must conceive of protoplasm to have certain kinds of properties which impose limitations on evolutionary development. These limitations apply particularly to chemical morphology. They also apply to behavioral developments as far as they are limited by morphological and genetic factors.

Reversibility

Peirce's definition of synechism, as given in Baldwin's (1902) dictionary, is discussed in chapter 2. In the following pages, the term will be used in a more general sense to apply to any guiding principle of ordered continuity in the realms of morphology and behavior. It is then not restricted by any purely logical or metaphysical definition. The terms continuity, reversibility or invariance are used, sometimes loosely, to convey similar meanings related to the constraints of morphological and behavioral evolution. A number of philosophers, metaphysicians and scientists may be cited in this context.

In the 17th century, Spinoza wrote in proposition VI, part III, of the Ethic, 'Each thing, insofar as it is in itself endeavors to persevere in its being.' Proposition IV, part III, states, 'A thing cannot be destroyed except by an outside source.' These 'geometrical' propositions in the language of modern physiology may be considered as principles of homeostasis or of self-identity, as later understood by the followers of Claude Bernard. If, for Spinoza's use of the word 'thing', we substitute Henderson's concept of 'physicochemical system', we arrive at the modern concept of homeostasis as applied to living organisms.

In the realm of purely physicochemical systems, we find a related principle of invariance or reversibility in Carnot's principle. This may be stated in the form: 'No work can be obtained from an invariant physicochemical system.' This principle also has biological applications to Bernard's principle of the constancy of the internal environment (milieu intérieur). The principle of homeostasis as applied by *Cannon* (1932), is somewhat different from that of *Henderson* (1928). Within the appropriate conditions of constraint, the physiological and physicochemical approaches are equally valid and generally applicable (*Joseph,* 1971b).

From the point of view of pure mathematics, *Caratheodory* (1909) has arrived at a statement of *Carnot*'s principle, which may be given in the form: 'There are inaccessible states in the neighborhood of any given state.' By the use of the Pfaffian differential equations of mathematics, this proposition may be obtained from invariant line integrals which yield the conditions for conservation of reversibility in any physicochemical system (*Born,* 1948). Thus, *Caratheodory*'s theorem for conservative line integrals, equivalent to *Carnot*'s principle, has been applied as a principle of homeostatic reversibility and agrees with modern statements of homeostasis *(Cannon, Henderson)* in affirming principles of invariance, continuity and constraint in 'each thing insofar as it is in itself'. *Peirce*'s concept of synechism is one of very wide generality. It becomes a highly generalized mathematical principle in *Caratheodory*'s abstract statement of *Carnot*'s principle.

In the course of his studies of physicochemical equilibrium in biological systems, *Henderson* (1928) became interested in the purely mathematical applications of *d'Ocagne* and Cartesian nomograms to physiological problems. These problems were also formulated in relation to *Gibbs' Equilibrium of Heterogeneous Substances.* From the purely mathematical point of view, *Henderson* found analogies to the principles of equilibrium and homeostasis in human sociology (*Pareto,* 1935). These generalized physicochemical and thermodynamic principles, from the point of view of mathematics, were found to be applicable to human societies (*Henderson,* 1935, 1970). Societies as a whole conform in behavior to the general principles of synechism and tychism. *Pareto*'s method, according to *Henderson,* 'is an application of the logical method that has been found useful in all physical sciences when complex situations involving many variables in a state of mutual dependence are described'.

Social systems do not conform to simple dualistic descriptions in terms of cause and effect, such as those that are adequate for the simplest kinds of mechanism. They can be described only by methods previously found necessary for dynamic, thermodynamic, physiological and economic systems. In such systems, it is necessary to analyze the mutual interdependence of many variables. Such systems are subject to many conditions of constraint. As in thermodynamics or in physiology, these conditions restrict the number of independent variables or degrees of freedom. In physicochemical systems or biological organisms, the degrees of freedom are determined by physicochemical state, the nature of protoplasm, and the chemical morphology. In human societies, the behavior is largely determined by legal and economic systems, as predetermined by historical conditions and conventional behavior.

After a lengthy section in which *Pareto* brings in mathematical considerations of systems involving the mutual interdependence of many factors in society, including non-logical forms of human behavior, he arrives at the demonstration of the importance of the *sentiments*. These are to be methodically studied in relation to any social system, are not directly observed, but are assumed or inferred. Their presence is often obscured by verbal usage and by non-logical forms of expression. The sentiments are not manifested singly and directly but as *aggregates*. No intellectual error is more common than to confuse the sentiments or the aggregates with rationality, when in reality human behavior is impregnated with non-rational elements. All students of humanity from the philosophers to the psychoanalysts have given witness to the importance of the irrational side of human nature. It was certainly known to the Greeks in the tragedies of Sophocles and Aeschylus, and in the comedies of Aristophanes. Life in the 20th century continues to manifest the tragicomic spirit.

Pareto continues his treatment of the sentiments and aggregates by developing the concepts of *residues* and *derivatives*. The words duty and justice refer to certain important residues. In the course of historical development, they gave rise to derivations such as theology and law. *Pareto* has attempted a classification of the residues. His classes include (1) residues of combinations and (2) persistence of aggregates. Persistent aggregates would account for the stable, invariant elements of human behavior in any and all human societies. They are the invariants, or what could be called the principle of synechism, continuity or cohesiveness. The residues of combinations would then include the factors of inventiveness, risk-taking, and adventurism. These would account for evolutionary changes and for periods of progress or decline. Rise and fall of the sentiments or aggregates would then depend on the rather fortuitous nature of these combinations. Social change then depends on the principle of tychism, just as continuity depends on synechism. In all societies, human or animal, homogeneous or heterogeneous, protoplasm must have inherent properties of invariance and homeostasis. But to survive, it must combine this element of stability with the elements of adaptability, risk-taking, and adventure. This involves the residues of courage and heroism.

Irreversibility

Irreversible changes in human behavior can often be traced to conspicuous individuals such as *Columbus* or *Newton*. The discovery of America, whether attributed to *Columbus* or to obscure Vikings of an earlier period, was an example of great courage, adventure and intelligent risk-taking and foresight. It led to a great historical break with the past and to the infusion of new human protoplasm and genes into the new continent. Not less important was the spread of the spirit of geographical discovery and adventure in all other parts of the earth, including all the oceans, continents and mountain ranges. Within the past century, geographical discovery has been extended to all the polar regions, and at present the only acceptable limits are beyond the biosphere and beyond the stratosphere.

The Newtonian adventure was established on the basis of the earlier achievements of such courageous predecessors as *Copernicus, Galileo* and *Kepler.* Thus 'tychism' established the grounds for the successful 17th century revolt against medieval Aristotelianism and for the rise of the experimental sciences. Since the results of this revolt have been on the whole beneficial, the resulting progress has been judged advantageous. In this instance, the residue of combinations has triumphed over the persistence of aggregates. There are no reputable medievalists in science, and in the advanced countries of the west, there are very few medievalists in the fields of law, justice or morality. The sentiments and persistent aggregates have changed, perceptibly and irreversibly. Only the residues based on the nature of protoplasm and the genes remain invariant: the needs of nutrition, reproduction, defense, and the demands of the five senses for pleasurable satisfaction. Thus tychism has won victories, but never at the expense of synechism, which barring cosmic catastrophe, must remain invincible.

While *Columbus* and *Newton* must forever remain heroic in expanding the boundaries of human behavior, lesser heroes and heroines should not be forgotten. Many of these will remain obscure and anonymous, like the first man who ate an oyster. Sir *Walter Raleigh* is remembered as the man who brought tobacco from Virginia to the Old World, thus introducing an irreversible change in human behavior. The first woman who smoked cigarettes in public, although less celebrated than *Raleigh,* was also responsible for an irreversible process. In the residue of combinations, she led a successful attack on the persistence of aggregates. The novelist *George Sand* (a friend of *Chopin*), was a cigar smoker, but her revolt was less successful. It may be too early to pronounce her permanent defeat.

In the animal kingdom, there are also adventurous heroes of tychism. Among these must be included the first brown rat that crossed the Volga. According to *Bourlière* (1953), this event occurred in 1727; it led to a continuous migration that resulted in the conquest of the sewers, wharves, cellars and refuse heaps of most of the human world. Successive outposts of this migration are said to have been Germany (1750), Paris (1753), Norway (1762), Spain (1800), Switzerland (1809). It entered the United States by sea in 1775, and appeared in Wyoming and Montana in the 20th century.

Although closely related by phylogeny, the brown rat differs significantly from the black rat in its behavior (ethology). The increase of human populations, with their multiplicity of cities, seaports, food supplies and wastes, has worked to the advantage of the brown rat. It is more aquatic than the black variety, and shows a number of other ecological advantages. Thus in the form of human population growth and ethological behavior, tychism has favored the survival and population growth of the brown rat. It also operates in the form of geographical distribution and ethology of competing forms, such as the black rat, the squirrels, and other rodents. In this way the behavior of rodent populations is related to that of the human population in every part of the earth. Problems of animal and human behavior cannot be individualized. They all depend on complex factors of chance and on the dynamics of all species of animals within the total population.

The list of species that have contributed to the great adventure of animal evolution might be expanded indefinitely. It would include the bats and flying squirrels – mammals that have developed aerial modes of locomotion. The first vertebrate to have emerged from ponds and swamps to adopt a terrestrial habitat was the lungfish (*Smith,* 1935). Birds that make seasonal migrations from the arctic to the antarctic circle must have descended from very adventurous ancestors. The great diversification of animal morphology and behavior that occurred over geological ages is due to the long-range operations of chance. It is an irreversible process. In no sense does it contradict the principles of synechism, reversibility, or conditions of biological invariance and constraint. These principles conform to the first and second laws of thermodynamics.

The principle of biological diversity across all the species is in no way restricted by the principles of continuity that operate over comparable periods on the geological time scale. Even in the most highly developed forms (vie constante), the second law (*Carnot*'s principle) requires the inaccessibility of all states that are not determined phylogenetically and ontogenetically by the laws of statistical probability. This, I think, is the only way to eliminate teleological or supernatural explanations from scientific thought. It cannot be done by the sophistry of identifying the second law with 'flat earth science'. This seems to be *Koestler*'s method in his book *The Ghost in the Machine* (1967). Being unable to reconcile the great diversity and development of highly ordered behavior in human and animal biology with *Carnot*'s principle, he seems ready to accept the idea that organic evolution is incompatible with thermodynamics. This seems to commit him to the idea that the second law, or the principle of increasing disorder in the universe, is incompatible with increasing order in the realm of human and animal ethology. *Koestler* implicitly assumes that the principles of thermodynamics apply only to closed or isolated systems that are not at equilibrium, and that they cannot be applied to biological 'open systems'.

In *Comparative Physical Biology* (1973), I have attempted to show that this science must be based on the principle that the biosphere of the earth is not an unrestricted open system, but that there are conditions of constraint that result from a vast population of living plants and animals. The presence of these living organisms introduces restrictions that channel and constrain the flow of energy and entropy through the biosphere. These constraints do not apply to the other planets, where the transmission of solar radiant energy is unrestricted by the requirements of living beings. This was clearly understood by *Boltzmann* (1905) and by *Schroedinger* (1944).

The second law of thermodynamics is not to be construed as a simple principle of order or disorder. It is necessary always to apply the proper conditions of constraint, as defined by *Gibbs* (1875) in his development of the phase rule. This is applicable without restriction to many kinds of open and closed systems, but only when the conditions of constraint are exactly defined. To limit thermodynamics or the phase rule to closed systems is to limit the application of physical chemistry to biology (*Joseph,* 1971b). As I have shown earlier, biological organisms are *restricted* open systems. This applies also to entire human societies, in which the essence of the problem is to define the conditions of constraint in the open system. This is necessary to establish the basic universe of discourse on which logical classification must depend. Otherwise, generalizations as to order, disorder and evolutionary development become merely sophisms. The second law of thermodynamics is properly to be regarded as a principle of irreversibility (tychism). Many kinds of irreversibility are mutually compatible, including progressive development of the principle of homeostasis. This is the sense in which I believe *Bernard* conceived of the evolutionary development of vie constante, which embraces all the forms of maximal order, invariance, and internal constraints. In this sense, the evolution from lower to higher forms of life implies an increase of order. This is the condition of freedom and independence in man, that enables him to transfer order to the external world.

Development of the Earth

Modern scientific theories as to the origin of the earth may be dated from the publication by *Laplace* of his *The System of the World,* which appeared late in the 18th century. This gave life to the 'nebular hypothesis', according to which the solar system was at one time part of a diffuse rotating nebula. As it cooled, the system contracted and spun with increasing velocity. During this process, the planets with their satellites were flung apart by centrifugal forces into elliptical orbits with the sun as a focus of each ellipse. The hypothesis accounts for the fact that the planets revolve around the sun in the same direction, and that in the majority of cases, the rotations occur in the same direction. Such an origin of the planetary system implies a cooling process. The temperature of the earth now varies from less than 273 $^{\circ}$K in the polar regions to about 300 $^{\circ}$K in the regions most favorable to human life, such as the temperate zones of Europe and America. The temperature of the sun is much higher, of the order of 6,000 °K at the surface. The cooling process in the case of the earth occurred over a period of about 4 billion years. During most of this period, conditions would have been unfavorable to the development of living forms.

Since the 18th century, many modifications of the nebular hypothesis have been found necessary. For one thing, later observers found by the study of the heavens that the motion of such systems as the observable spiral nebulae was not of the nature that could lead to condensation of spherical bodies like our planets. Considerations as to the angular momenta in such a system as our solar system require fundamental modifications of *Laplace*'s hypothesis. These modifications, which varied in many details, had one thing in common — a *catastrophe* in the sun. All the stars in the heavens are moving at tremendous speeds. The sun moves with a velocity of about 12 mi/sec. In the course of time, a cosmic catastrophe resulting from a collision between two stars would occur. It has been established that such an event could happen to only 1 of a 100 million stars. But it is now thought that the origin of the solar system was produced by such a collision about 4 billion years ago, and that it was followed by processes of cooling and condensation that have been going on ever since.

Such processes of condensation and cooling would imply a continuous and progressive decrease of entropy on the earth. It would eventuate in the formation of liquid water and of ice from water vapor. This would begin at the time when the temperature of the earth had cooled, respectively, to 373 °K and 273 °K. Liquids and solids are known as condensed states of matter, as referred to the vapor state. This applies to all substances. Higher temperatures of the earth would have favored reactions of the elements in the gas state to form compounds of the nature of methane and other hydrocarbons and of ammonia and its derivatives. Continuing development of geophysical conditions would have favored the formation of many organic substances from the elements: nitrogen, oxygen, hydrogen and carbon. This could be thought of as a progressive process of polymerization, which would favor the evolution of heterogeneous phases.

In the evolution of life, a critical point would have occurred when photosynthesis was developed. We know of this mainly as the formation of high polymers from carbon dioxide and water. Both reactions involve a decrease of entropy associated with the growth and development of protoplasm. This process could then be associated with the progressive condensation or polymerization of matter that became possible when the earth had reached a sufficiently low temperature. The condensation process evolves through the formation of liquids and solids from the earliest gaseous states. It continues progressively through the formation of polymers of high molecular weight, and eventually becomes part of a well-ordered process of continuous biological development. The initial irreversible cosmic catastrophe was a break in the invariance and continuity of the cosmos, at least in the region of what is now our solar system.

The development now continued under geochemical conditions established by the initial state. The cooling of the earth is a consequence of the second law of thermodynamics. Energy was transferred to outer space, and entropy of the whole increased. However, after the development of living forms not all of the solar energy went to entropy. Some of it went into the biosphere, where living forms were continually evolving with the development of condensed macro-molecular polymers. This development required such processes as the condensation of carbon dioxide, water and nitrogenous substances to polymers such as proteins and starches. The entire evolutionary process then can be summarized as the progressive formation of well-ordered macromolecular systems from poorly-ordered states of elementary substances. This implies a great diversification of protoplasmic morphology and ethology. It also implies an origin of sensations and perceptions.

Irritability

When a minute speck of protoplasm such as an amoeba or a slime mould is irritated at the surface, it shows the property of *feeling*. Since the 17th century, this property has been known as *irritability*. In primitive form it is the prototype of the property that is exhibited by a stimulated nerve ending. At the point in the amoeba where the irritability occurs, a motion of the cytoplasm is observed. A disordered process of liquefaction begins to spread through the protoplasm, with all the appearance of a change of state from a solid or semi-solid to a liquid state of aggregation. This process is not organized, as in the vertebrate nerve cell, to propagate the stimulus toward a definite terminus. It has the character of a disorganized change of behavior of an amorphous substance leading to motion away from the source of the sensation. The motion resembles a kind of viscous flow. It would be reasonable to assume, as in other kinds of irritability or sensation at the cellular level, that Brownian movement would be increased by the process of viscous flow, and that there would be changes of dielectric energy and dielectric constant.

These points may be tested from the following results of *Bruce and Marshall* (1965). In the amoeba *Chaos chaos*, an intracellular electrical potential of $-85\,mV$ was observed in an external medium that contained sodium at a concentration of 47 mmole/l. The cytoplasmic concentration $(c_{Na})''$ was 13 mM per kg water. From these results, the electrical potential may be converted to $\Delta\mu_{Na}$ in kilocalories by the conversion factor, 1 kcal = 43.4 mV (chapter 5). Then the value is 1.95 kcal. By means of equation 5.11, the value of $\Delta\mu_{Na}{}^0$ is estimated as 2.50 kcal, when r_{Na} is taken as 13/47. The corresponding value of the dielectric constant is 41, and the value of the dielectric energy is 2.5 cal/kg water. For cephalopod axons, the dielectric energy is of the order of 100 cal/kg water, and in human brain the value is about 50 cal (table VII). It is evident that the energy requirements of the amoeba for sensory perception or for its motility is of a lower order of magnitude than the requirements of nerve tissue of higher organisms. It is interesting, however, that the resting potential (-85 mV) is of the same order of magnitude as that found in mammalian cells.

The energy requirements of an amoeba or a slime mold are certainly very meager. Estimating the volume of the amoeba as that of a sphere of radius 200 μm, the water content is of the order of 3×10^{-5} g, or 3×10^{-8} kg. If the dielectric energy is 2.5 cal/kg water, the value amounts to 7.5×10^{-8} cal. This would be the maximal energy available in any change of state yielding the total dielectric energy. A more likely figure for amoeboid movement might be of the order of 10^{-10} cal for a single amoeba. Comparing this with the figure for mammalian skeletal muscle, it is recalled that the latter figure is of the order of 85 cal (350 J) per kilogram water. This amounts to about 90 J (9 kg m) for the arm muscles of an adult man (*Hill*, 1944, 1951; *Joseph*, 1971a, b; 1973). The dielectric energy of the human arm muscles exceeds that of an amoeba by a factor of the order of 10^{12} (1 trillion). This would seem to be a measure of the extreme diversity of work capacity or configurational free energy between man and protozoa.

In the case of a human athlete, performance at the level of Olympic competition requires the ability to make the full use of his muscular energies. This implies the full development of his neuromuscular skills, a process that may require many years of training. This is an adventure in tychism. In the untrained athlete, one cannot say that:

behavior = morphology.

At the human level, one must say:

behavior = morphology + training.

For the amoeba or the slime mold, it would be true that behavior is equal to morphology. These creatures have probably no learning capacity. At the lowest levels of animal life, mentality and reasoning ability reach the absolute lower limit for protoplasm in any form. Then it would be proper to say that there is no difference between the primary property (morphology) and the secondary property (behavior). At this limit *Occam*'s razor is completely applicable, and behavior can be eliminated as an independent entity. If one entity is sufficient, the second is unnecessary. But at this point, it is also possible to say that morphology is equal to dielectric energy. It is unnecessary to introduce energy when it is synonymous with cell morphology. At this point, existence and behavior become identical, and Cartesian dualism completely vanishes. There is no necessary distinction between primary and secondary properties or between organism and environment. At this unicellular level, the element of adventurism likewise vanishes. There is no ego to be detached from the non-ego.

Viscous Flow in Protoplasm

Unicellular organisms are not structured to transmit dielectric energy along organized paths as in the sensory or motor neurones of higher organisms. There are no reflex arcs or organized centers to produce any kind of organized or purposeful behavior. Irritability is limited to a change of state of protoplasm, which without more exact definitions may be termed depolymerization, dis-aggregation, or simply change of hydration. This change of state may imply a decrease of dielectric energy of the order of 2.5 cal/kg cell water. Unlike the peripheral or central neurones of the nervous systems of vertebrates, unicellular organisms are not developed morphologically to distribute this energy in any integrated processes. The depolymerization process may spread through many parts of the protoplasm, but the spreading would seem to partake of the nature of viscous flow as governed by laws of chance and statistical probability rather than by any principle of invariance or continuity.

The principles of diffusion and viscous flow of macromolecular polymers were worked out in a series of papers by *Einstein* on the Brownian movement (1905—1908). In systems in which the principles of ordinary hydrodynamics may be applied to large spheres moving in a liquid of viscosity, η, the following equation is valid:

$$A = 6 \pi \rho \eta$$

where ρ is the radius of the sphere, A is a constant which *Einstein* has called the frictional resistance of the molecule. This is related in the following way to the diffusion constant by:

$$D = \frac{RT}{NA}$$

This is equivalent to equation 4.6, as stated in chapter 4, dealing with Brownian movement and diffusion. In that chapter, it was stated that, although Brownian movement, diffusion and spreading phenomena must be operative in mammalian cells and connective tissues, they would be of a subordinate nature in physiology.

In mammalian physiology, dielectric energy would be necessarily channeled through well-ordered and highly structured systems of neurones, synapses and muscle fibers. Processes of spreading and diffusion would be, to a large extent, limited to the amorphous ground substance of connective tissue, the structure of which depends to a large extent on growth, aging and endocrine control. Very fast reactions such as nerve impulses and muscular contraction are organized on the basis of submicroscopic morphology and are largely independent of the random motions of dispersed particles that lead to the phenomena of diffusion.

Diffusion and spreading are poorly-ordered phenomena of chance, rather than well-ordered phenomena that can occur only in highly organized invertebrates or in vertebrates at higher states of evolution. Spreading phenomena operate at all levels of evolution and in mammals would be important in determining the rates of equilibration of narcotics and anesthetics with sensory nerves (chapter 8). Very fast reactions such as the perception of sound or visual images imply the almost instantaneous transmission of dielectric energy through nerve endings, sensory endings and synapses of the central nervous system. These processes imply extremely rapid oscillations of dielectric constant and dielectric energy, and very rapid changes of state of protoplasmic structure. These highly ordered transmissions of dielectric energy are characteristic only of higher forms of life. They differ in kind from the most primitive kinds of irritability and sensation that are found at the unorganized unicellular level. In higher organisms, phenomena of diffusion and viscous flow would be important only in the distribution of nutrients, metabolites, drugs and other physiologically active substances through the extracellular phases of connective tissues. Compared with the very high speed of normal neuromuscular processes of vertebrates, primitive processes of diffusion and Brownian movement are extremely slow and disorganized.

The Faculty of Reason

Like organisms, mechanisms may be purposeful, but they do not determine their own purposes. Like all other things, they endeavor to persevere in their own being. Unlike mechanisms, living organisms of the more highly evolved species are reasonable things that endeavor to preserve invariant internal conditions. To understand man, who stands at the pinnacle of reasoning organisms, we must inquire into the nature of nonmechanistic reason. According to *Whitehead* (1929), there are three functions of reason: to live, to live well, and to live better. To ask what a philosopher understands by living well or by living better is to ask what he means by human values. Since the time of *Plato* and *Aristotle,* all philosophers have been occupied with the *why* of things, or the rationality of their *being.*

In this the philosopher or metaphysician resembles the young child who insists on knowing the reason for everything he encounters. The great philosopher *Samuel Alexander* insists that this child-like attitude of 'natural piety' be retained as a foundation of metaphysics. This is the attitude of the poet *Wordsworth:*

'The child is father of the man,
And I could wish my days to be
Bound each to each by natural piety.'

The development of modern experimental science has been attended by an anti-metaphysical bias of the kind that adults acquire when they begin to suppress their child-like natural piety. Many scientists are proud of this rather hard-boiled attitude, which is expressed most strongly in the various schools of positivism.

In extreme forms an anti-metaphysical bias can take the form of an attitude of active irrationalism. In his *Essay on Metaphysics, Collingwood* (1938) has explained the nature of various kinds of anti-metaphysics, including the positivistic form. 20th century irrationalism, as we know, has been capable of vicious intolerance of many kinds. As *Collingwood* shows, a true definition of metaphysics requires the criticism of the absolute preconceptions of each special science. This was *Aristotle*'s understanding of the term. It would include the criticism not only of our preconceptions in the biological sciences, but also in ethology, the science of behavior.

According to various anti-positivistic trends in philosophy and ethics, such criticism could develop an understanding of the function of reason that would be appropriate to the needs of the coming centuries. At the human level, ethology certainly calls for an understanding of the mind and body in relation to the requirements of all human beings for the satisfaction of their physical and emotional needs. This should recognize a continual reconsideration of the absolute preconceptions of the biological sciences. In future times it may be imperative to reorganize many of the natural sciences in relation to the needs of human reason.

Bibliography

Adams, H.: The education of Henry Adams (Modern Library, New York 1900).

Alexander, S.: Space, time and deity (Dover, New York 1960).

Atwater, W.D.; Woods, C.D., and Benedict, F.G.: US Department of Agriculture Bull. *44:* 51 (1897).

Bailey, P. and Bonin, G. von: The isocortex of man (University of Illinois Press, Urbana 1951).

Baldwin, J.M.: Dictionary of philosophy and psychology (Finch Press, 1902).

Balint, M.: Thrills and repression (Hogarth Press, London 1959).

Benedict, F.G. and Cathcart, C.J.: Muscular work (Carnegie Institute, Washington 1913).

Bernard, C.: Leçons sur les phénomènes de la vie communs aux animaux et aux végétaux (Baillière, Paris 1878).

Bernard, C.: Experimental medicine (Henry Schumann, 1927).

Bernard, C.: (1865): The cahier rouge (Schenkmann, Cambridge 1967).

Boedtker, H. and Simmons, H.S.: Preparation and characterization of essentially uniform tobacco mosaic virus particles. J. Am. chem. Soc. *80:* 2550–2556 (1958).

Boltzmann, H.: Populäre Schriften (Leipzig, 1905).

Bondareff, H.: Morphology of connective tissue ground substance with particular reference to fibrillogenesis and aging. Gerontologia *1:* 222–233 (1957).

Born, M.: Volumes and heats of hydration of ions. Hoppe-Seyler's Z. physiol. Chem. *1:* 45–48 (1920).

Born, M.: Natural philosophy of cause and chance (Oxford University Press, London 1948).

Bourlière, F.: The natural history of mammals (Knopf, New York 1953).

Brodmann, E.: Neue Ergebnisse über die vergleichende histologische Lokalisation der Grosshirnrinde mit besonderer Berücksichtigung des Stirnhirns. Anat. Anz. (1912).

Brown, C.G. and Sherrington, C.S.: On the irritability of a cortical point. Proc. R. Soc. B *85:* 250–277 (1906).

Bruce, D.M. and Marshall, J.M., jr.: Some ionic and bioelectric properties of the amoeba *Chaos chaos.* J. gen. Physiol. *49:* 131–178 (1965).

Brul, E.L. du: Evolution of the speech apparatus (Thomas, Springfield 1958).

Cannon, W.B.: The wisdom of the body (Norton, New York 1929).

Cantor, G.: Contributions to the founding of the theory of transfinite numbers. Part I. Math. Ann. *46:* 481–512 (1895).

Cantor, G.: Contributions to the founding of the theory of transfinite numbers. Part II. Math. Ann. *49:* 212–246 (1897).

Caratheodory, C.: Investigations of the fundamentals of thermodynamics. Math. Ann *67:* 355–386 (1909).

Catchpole, H.R. and Joseph, N.R.: Muscular work and athletic records; in Nelson and Morehouse, Biomechanics IV (University Park Press, Baltimore 1974).

Chase, W.H.: Archs Path. *67:* 525 (1959).

Christensen, J.J.; Hill, J.O., and Izatt, R.M.: Ion binding by synthetic macrocyclic compounds. Science *174:* 459–467 (1971).

Cohn, E.J. and Edsall, J.T.: Proteins, amino acids and peptides (Reinhold, New York 1943).

Collingwood, W.B.: An essay on metaphysics (Regnery, Chicago 1938).

Darlington, C.D.: Evolution of man and society (Simon & Schuster, New York 1969).

Debye, P.: Molecular weights of proteins from light scattering. J. phys. Colloid Chem. *51:* 18–32 (1947).

Dennis, J.B.: Archs Path. *69:* 533 (1959).

Dickerson, J.W.T. and Widdowson, E.M.: Chemical changes in skeletal muscle during development. Biochem. J. *74:* 247–257 (1960).

Donnan, F.G.: Theory of membrane equilibrium and membrane potentials in the presence of a non-dialyzable electrolyte. J. Electrochem. *17:* 572–581 (1911).

Ecanow, B.; Gold, B.E., and Tunkemann, F.: Application of physical chemical principles to the study of anxiety and depression. Psychiatry *21:* 121–127 (1972/1973).

Ecanow, B. and Klavans, B.I.: A physical chemical model of membranes in models of human neurological disease; 1st ed. (Excerpta Medica, Amsterdam 1974).

Einstein, A.: Investigations of the theory of the Brownian movement; transl. A.D. Cowper (Methuen, London 1926).

Engel, M.B.; Joseph, N.R.; Laskin, D.M., and Catchpole, H.R.: Am. J. Physiol. *201:* 621 (1961).

Erdos, T. and Snellman, O.: Biochim. biophys. Acta *2:* 647 (1948).

Fenn, W.B.: Electrolytes in muscle. Physiol. Rev. *16:* 956–977 (1935).

Ferguson, J.: The use of chemical potentials as indices of toxicity. Proc. R. Soc. B *127:* 387–409 (1939).

Flory, P.I.: Principles of polymer chemistry (Cornell University Press, Ithaca 1953).

Frey-Wyssling, A.: Submicroscopic morphology of protoplasm (Elsevier, Amsterdam 1953).

Fritsch, O. und Hitzig, E.: Über die elektrische Eregbarkeit des Grosshirns. Arch. Anat. Physiol. *12:* 171–233 (1870).

Gersh, I. and Catchpole, H.R.: The nature of ground substance of connective tissue. Perspect. Biol. Med. *3:* 292–342 (1960).

Gersh, I.; Isenberg, I.; Stephenson, J.L., and Bondareff, W.: Submicroscopic structure of frozen-dried liver specifically stained for electron microscopy. I. Techn. Anat. Rev. *128:* 91–111 (1957).

Gibbs, J.W. (1875): On the equilibrium of heterogeneous substances (Longmans Green, New York 1928).

Gibbs, J.W. (1901): Elementary principles in statistical mechanics (Longmans Green, New York 1928).

Gilson, E.: The unity of philosophical experience (Scribners, New York 1938).

Glisson, F.: Anatomia hepatica (Amsterdam 1650).

Goldschmidt, V.M.: Geochemical distribution of the elements. Skriften Norske Wicenskamps Akad. (Oslo). J. Mat. Natur. Kl. 8 (1926).

Havet, J.: Contributions a l'étude des systèmes nerveux des actinies. Cellule *18:* 285–418 (1901).

Head, H.: Speech and cerebral localization. Brain *46:* 355–526 (1923).

Heisenberg, W.: The physical concept of nature (Harcourt Brace, New York 1958).

Henderson, L.J.: The fitness of the environment (Macmillan, New York 1908).

Henderson, L.J.: Blood, a study in general physiology (Yale University Press, New Haven 1928).

Henderson, L.J.: On the social system (University of Chicago Press, Chicago 1970).

Helmholtz, H. von: Über die Erhaltung der Kräfte (Berlin 1847).

Helmholtz, H. von (1862): The sensation of tone as physiological foundation for the theory of music (Dover, New York 1954).

Herrick, C.J.: The evolution of human nature (Harper, New York 1953).

Hertwig, O. und Hertwig, H.: Das Nervensystem und die Sinnesorgane des Menschen (Vogel, Leipzig 1878).

Hill, A.V.: Living machinery (Bell, London 1944).

Hill, A.V.: The mechanics of voluntary muscle. Lancet *ii:* 942–951 (1951).

Hodgkin, A.L.: The ionic basis of electrical activity in nerve and muscle. Biol. Rev. *26:* 330–405 (1951).

Huxley, T.H.: The physical basis of life (Lay sermon, 1868); quoted in *Hardy* Collected papers (Cambridge University Press, London 1936).

Jolly, W.A.: The time relations of the knee-jerk and simple reflexes. Q. Jl exp. Physiol. *4:* 67–87 (1910).

Joseph, N.R.: Physical chemistry of aging (Karger, Basel 1971a).

Joseph, N.R.: Dependence of electrolyte balance on growth and development of cells and tissues; in *Elden* A treatise on skin (Wiley Interscience, New York 1971b).

Joseph, N.R.: Comparative physical biology (Karger, Basel 1973).

Joseph, N.R.: Physicochemical anthropology. Part I: Human behavioral structure (Karger, Basel 1978).

Joseph, N.R.; Engel, M.B., and Catchpole, H.R.: Distribution of sodium and potassium in certain cells and tissues. Nature, Lond. *191:* 1175–1178 (1961).

Katz, J.: Die mineralischen Bestandteile des Muskelfleisches. Pflügers Arch. ges. Physiol. *61:* 1 85 (1895).

Koenig, R.: Quelques expériences. Ann. Physik *69:* 626–660 (1884).

Koestler, A.: The ghost in the machine (Macmillan, New York 1967).

Laidler, K.J. and Pegis, C.: The influence of dielectric saturation on the thermodynamic properties of aqueous ions. Proc. R. Soc. A *221:* 80–92 (1957).

Levi-Strauss, C.: The savage mind (University of Chicago Press, Chicago 1970).

Lew, V.L.: Renewing the search for calcium pumps. Nature, Lond. *274:* 421–422 (1978).

Lewis, G.N. and Randall, M.: Thermodynamics (McGraw-Hill, New York 1961).

Loeb, J.: The dynamics of living matter (Columbia University Press, New York 1906).

Lorenz, K.: Studies in animal and human behavior, vol. 1 and 2 (Harvard University Press, Cambridge 1970–1972).

Macallum, A.B.: The inorganic composition of the blood in vertebrates and invertebrates and its origin. Proc. R. Soc., Lond. *82:* 602–624 (1910).

Macallum, A.B.: The paelochemistry of the body fluids and tissues. Physiol. Rev. *14:* 133–159 (1926).

Mayer, J.R.: Bemerkungen über die Kräfte der unbelebten Natur. Justus Liebigs Ann. Chem. *42:* 233 (1842).

McCulloch, W.S. and Pfeiffer, J.: Of digital computers called brains. Sci. Mon. *69:* 368–376 (1949).

Miller, D.C.: The science of musical sounds (Macmillan, New York 1916).

Morehouse, L.E. and Miller, A.T.: Physiology of exercise (Mosby, St. Louis 1967).

Naora, H.; Naora, H.; Izawa, W., and Alfrey, V.G.: Some observations on differences in composition between the nucleus and cytoplasm of the frog oocyte. Proc. natn. Acad. Sci. USA *48:* 853–859 (1962).

Oparin, A.: Origin of life (Macmillan, New York 1938).

Paget, R., Sir: Human speech (Kegan Paul, London 1963).

Pareto, V.: The mind and society (Harcourt Brace, New York 1935).

Parker, G.H.: The elementary nervous system (Lippincott, Philadelphia 1918).

Pauling, L.: The nature of the chemical bond (Cornell University Press, Ithaca 1944).

Peirce, C.S.: Man's glassy essence. The Monist (1892).

Peirce, C.S.: Deduction, induction and hypothesis. Popular Science Monthly, August (1878).

Peirce, C.S.: Chance, love and logic (Harcourt Brace, New York 1923).

Peirce, C.S.: in *Buchler* Philosophical writings (Dover, New York 1955).

Penfield, W. and Roberts, C.: Speech and brain mechanisms (Princeton University Press, Princeton 1955).

Penfield, W. and Rasmussen, T.: The cerebral cortex of man (Macmillan, New York 1959).

Rayleigh, Lord (J.W. Strutt): Phil. Mag. *11:* 107 (1871).

Rayleigh, Lord: Phil. Mag. *47:* (1899).

Robertson, J.D.: The inorganic composition of muscle. II. The abdominal flexor muscle of the lobster *Nephrops norvegicus.* J. exp. Biol. *38:* 707–731 (1961).

Rothschild, Lord and Barnes, H.: The inorganic composition of the sea urchin egg. J. exp. Biol. *38:* 530–541 (1953).

Schleiden, M.: Grundrisse der wissenschaftlichen Botanik (Leipzig 1842).

Schroedinger, E.: What is life? (Cambridge University Press, Cambridge 1944).

Schwann, T.: Mikroskopische Untersuchungen über die Übereinstimmungen in der Struktur und dem Wachstum der Tiere und Pflanzen (Berlin 1839).

Shaw, J.: Ionic regulation of the muscular fibers of *Carcinus maenas.* I. The electrolyte composition of single fibers. J. exp. Biol. *32:* 385–396 (1955).

Shaw, J.: Osmoregulation in the muscle fibers of *Carcinus maenas.* J. exp. Biol. *35:* 385–396 (1958a).

Shaw, J.: Further studies of ionic regulation in the tissue fibers of *Carcinus maenas.* J. exp. Biol. *35:* 920–929 (1958b).

Sherrington, C.S.: Observations on the scratch reflex in the spinal dog. J. Physiol., Lond. *34:* 1–50 (1906).

Sherrington, C.S.: The integrative action of the nervous system (Yale University Press, New Haven 1920).

Sherrington, C.S.: Man on his nature (Macmillan, New York 1940).

Sicher, H. and Brul, E.L. du: Oral anatomy (Mosby, St. Louis 1970).

Smith, H.W.: From fish to philosopher (Little, Brown, Boston 1935).

Smoluchowski, M.: Phys. Z. *13:* 1–60 (1912).

Snellman, O.: Scand. J. clin. Lab. Invest. *2:* 248–251 (1950).

Snellman, O.: On actomyosin as protein complex. Biochim. biophys. Acta *5:* 56–58 (1952).

Snellman, O. and Erdos, T.: Biochim. biophys. Acta *2:* 650–660 (1948).

Spaulding, E.G.: The new rationalism (Holt, New York 1918).

Starling, E.H.: The law of the heart. Linacre Lecture (Longmans, London 1918).

Steinbach, H.B. and Spiegelman, S.G.: Chemical and concentration potentials in the giant fibers of spinal nerve. J. cell. comp. Physiol. *15:* 373 (1940).

Steinbach, H.B. and Spiegelman, S.G.: The sodium and potassium behavior of squid neuro axioplasm. J. cell. comp. Physiol. *22:* 182 (1943).

Svedberg, T.: Colloid chemistry (Chemical Catalog, New York 1928).

Svedberg, T.: Über die Ergebnisse der Ultracentrifugation und Diffusion für die Eiweiss-chemie. Kolloid Z. *85:* 119 (1938).

Thompson, D'Arcy W.: On growth and form (Macmillan, New York 1944).

Waddington, C.H.: The ethical animal (University of Chicago Press. Chicago 1967).

Westgren, A.: Arkiv. fur Matematik., Stockh. *11:* No. 8 (1916).

Wheeler, W.M.: Emergent evolution (Norton, New York 1928).

Whitehead, A.N.: Science and the modern world (Macmillan, New York 1925).

Whitehead, A.N.: Process and reality (Macmillan, New York 1929).

Widdowson, E.W. and Dickerson, J.W.T.: Chemical composition of the body; in *Comar and Bronner* Mineral metabolism, vol. *2a* (Academic Press, New York 1960).

Wiener, N.: Cybernetics (Technology Press, Cambridge 1948).

Wilmer, E.A.: Changes in structural components of human body from six lunar months to maturity. Proc. Soc. exp. Biol. Med. *43:* 545 (1940).

Willstatter, R.: Naturwissenschaften *15:* 585 (1927).

Willstatter, R.: Untersuchungen über Enzyme (Berlin 1928).

Young, J.Z.: The memory system of the brain (University of California Press, Berkeley 1966).

Subject Index